DOCTRINE UNDER TRIAL

Recent Titles in
Contributions in Military Studies

The Making of a Professional: Manton S. Eddy, USA
Henry Gerard Phillips

Jérôme Bonaparte: The War Years, 1800–1815
Glenn J. Lamar

Toward a Revolution in Military Affairs?: Defense and Security at the Dawn of the Twenty-First Century
Thierry Gongora and Harald von Riekhoff, editors

Rolling the Iron Dice: Historical Analogies and Decisions to Use Military Force in Regional Contingencies
Scot Macdonald

To Acknowledge a War: The Korean War in American Memory
Paul M. Edwards

Implosion: Downsizing the U.S. Military, 1987–2015
Bart Brasher

From Ice-Breaker to Missile Boat: The Evolution of Israel's Naval Strategy
Moshe Tzalel

Creating an American Lake: United States Imperialism and Strategic Security in the Pacific Basin, 1945–1947
Hal M. Friedman

Native vs. Settler: Ethnic Conflict in Israel/Palestine, Northern Ireland, and South Africa
Thomas G. Mitchell

Battling for Bombers: The U.S. Air Force Fights for Its Modern Strategic Aircraft Programs
Frank P. Donnini

The Formative Influences, Theories, and Campaigns of the Archduke Carl of Austria
Lee Eysturlid

Great Captains of Antiquity
Richard A. Gabriel

DOCTRINE UNDER TRIAL

American Artillery Employment in World War I

Mark E. Grotelueschen

Foreword by Tim Travers

Contributions in Military Studies, Number 203

GREENWOOD PRESS
Westport, Connecticut • London

Library of Congress Cataloging-in-Publication Data

Groteluesschen, Mark E., 1969–
Doctrine under trial : American artillery employment in World War I / by Mark E. Grotelueschen ; foreword by Tim Travers.
p. cm.—(Contributions in military studies, ISSN 0883–6884 ; no. 203)
Includes bibliographical references and index.
ISBN 0–313–31171–4 (alk. paper)
1. United States. Army—Army, 1st—Artillery. 2. World War, 1914–1918—United States. 3. United States. Army—Artillery operations, American. 4. World War, 1914–1918—Campaigns—France. 5. Artillery drill and tactics—History—20th century. 6. Military art and science—United States—History—20th century. I. Title. II. Series.
D570.271st.G76 2001
940.4'1273—dc21 00–029423

British Library Cataloguing in Publication Data is available.

Library of Congress Catalog Card Number: 00–029423
ISBN: 0–313–31171–4
ISSN: 0883–6884

First published in 2001

Greenwood Press, 88 Post Road West, Westport, CT 06881
An imprint of Greenwood Publishing Group, Inc.
www.greenwood.com

Printed in the United States of America

The paper used in this book complies with the Permanent Paper Standard issued by the National Information Standards Organization (Z39.48–1984).

10 9 8 7 6 5 4 3 2 1

For Alison, my wonderful wife

Contents

Illustrations

MAPS

FIGURES

Foreword

Mark Grotelueschen has written a clever book, in which an initial investigation of the artillery methods of the 2nd U.S. Division in World War I turns into a larger analysis of the evolution of American doctrine. At the same time, Grotelueschen takes on the two main historical schools of thought about the American role in World War I, the Pershing advocates, and the revisionists, and finds a more accurate position in between these two schools.

The connection between artillery methods and doctrine emerges as Pershing and his General Headquarters (GHQ) pursue their well-known "open warfare" ideas, while the divisional commander of the 2nd U.S. Division and his commander of artillery soon find out that what their French trainers have been telling them is correct. This is that open warfare was not likely to work even in the more mobile environment of 1918, while carefully prepared set-piece battles were the key to success. Eventually, after hard won and bloody experience, the 2nd U.S. Division chose to adapt to the set-piece battle, although Pershing and his GHQ managed to keep their doctrine alive by insisting on deep objectives and open warfare exploitation after the set-piece battle ended. Did this finally work? Grotelueschen comes to his own conclusions about that question.

Grotelueschen's story of the struggle between competing doctrines also illuminates the nature of innovation. Where do changes and new ideas come from? Conversely, why do some commanders fail to change and adapt? This book is valuable also for comparative purposes. The reader who is familiar with the French, British and German commanders and their armies can find some interesting parallels in 1918. For the French, Foch and his prewar aggressive attack doctrine, like Pershing's, does not seem to change much, despite the brutal experience of horrific casualties and mutiny in the French army. On the other hand, Petain does conform to the necessities of firepower, as does the 2nd U.S. Division. For the British, Haig did not really understand technology and, like Pershing, usually insisted on objectives that were too deep, and kept offensives going for too long. But Haig's Army, Corps and Divisional commanders tended to develop their own methods, and did generally adapt to the set-piece battle, and succeeded in 1918, just like the 2nd U.S. Division's commander and his chief of artillery. Innovation seemed to come more easily to the German army, where Hindenburg and Ludendorff accepted new styles of warfare that initially resulted in more or less open warfare. Pershing was naturally attracted to this style of attack, but various German limitations and a stout Allied defense prevented the German army from succeeding in the spring of 1918.

Grotelueschen ends his book by revisiting the Pershing versus revisionist debate, and then argues that the experience of World War I contributed to American arguments over two fundamental approaches to warfare in the inter-war period.

Readers of this book will find a wide range of issues that emerge from the harsh reality of the trenches, the problems of firepower, the difficulty of preventing high casualties, and the important conflict over doctrine. There are also descriptions of battles, technology, and the first-hand encounter of raw troops with an experienced enemy. In the end, the wonder is not that the 2nd U.S. Division and the American Expeditionary Force had problems in trying to solve a new and puzzling style of warfare, but that they adapted so well in such a short time.

Tim Travers
University of Calgary

Acknowledgments

I am grateful to the many persons who offered assistance, advice and encouragement during the preparation of this book. Professor Timothy H. E. Travers, of the University of Calgary, provided much-needed guidance and advice throughout the enterprise, from the refinement of the topic to the final revisions. Also in the Department of History at the University of Calgary, Professor Holger Herwig suggested some important changes. Dr. Dennis Showalter, Professor of History at Colorado College and a visiting professor at the U.S. Air Force (USAF) Academy, offered extensive feedback on the manuscript and encouraged me to publish it. Numerous other friends took time away from their busy schedules to provide technical assistance and moral encouragement. Mélanie Méthot did both, most notably by translating various French Army reports into English, and Yvonne Youngberg, David Campbell, and Wesley Gustavson offered much-needed advice and encouragement.

My research efforts were greatly aided by the valuable assistance offered by John J. Slonaker and the staff at the U.S. Army Military History Institute, Carlisle Barracks, PA. At the National Archives in Washington, D.C., Mitchell Yockelson demonstrated a mastery of the massive collection of American Expeditionary Forces (AEF) records, and in many cases, pointed me to important files and documents that I did not know existed. Andrew Woods provided assistance at the McCormick Research Center of the First Division Museum in Wheaton, Illinois. Finally, Steve Novak, a friend and fellow officer in the United States Air Force, graciously opened his home to me for my lengthy stay in Washington while researching; I thank him for his hospitality, which was above and beyond the call of duty.

I want also to thank my parents, brothers, and sisters for all of the support, encouragement, and prayers they offered on my behalf. Of even greater importance to the successful completion of this project was the love and patience displayed by my wonderful wife, Alison. To Abigail, Grant, and Caleb, I offer thanks for continually showering their father with love, even after he spent too much time at the computer.

To all of the above, I offer my thanks and appreciation. Without such assistance, this task could not have been completed. Nevertheless, I am solely responsible for any and all mistakes, errors, or faulty conclusions contained in the work. Finally, as always, in the words of J. S. Bach, "Soli Deo Gloria."

Abbreviations

AEF	American Expeditionary Forces
BEF	British Expeditionary Force
FA	Field Artillery
GHQ	General Headquarters
HQ	Headquarters
PC	Post of Command

Introduction

> The war has repeatedly demonstrated that any position, however strong, can be neutralized and captured, with a sufficiently powerful fire, and that invariably the lack of sufficient superiority of firepower is paid for in losses to the Infantry and even in failure.[1]
>
> Major General Charles P. Summerall
> Commanding General, V Corps

When General Summerall referred to "firepower," he meant the kind of heavy fire support that only artillery could provide. Having commanded an artillery brigade, infantry division, and army corps in the American Expeditionary Forces (AEF) during World War I, Summerall had learned that a unit's artillery not only influenced the number of casualties sustained in any battle, but was often the difference between success and failure. Few historians, if any, would deny the crucial role played by artillery forces on the Western Front throughout World War I. Artillery, along with the machine gun, helped establish trench warfare in late 1914, was a major factor in maintaining trench warfare from 1915 through 1917, and in 1918, through developments in materiel, tactics, and techniques, played a primary role in returning a measure of mobility to the battlefield.

While numerous historians have written about the artillery theories, tactics, and operations of the other major World War I powers, no historian has focused on the role played by artillery in the operations of the AEF.[2] Most of the works that address the subject allocate a single, short chapter to AEF artillery as part of a general history on American artillery from the eighteenth century to the present day.[3] Other books that mention AEF artillery employment deal with the combat experiences of the entire AEF throughout the war. While these surveys occasionally mention specific details, such as the number of guns used in a given attack, or the length of a particular artillery bombardment, none offers a detailed examination of the organization, combat doctrine, training, tactics, or techniques used by the American artillery forces. Considering the crucial role artillery played in World War I warfare, this omission is surprising. Most histories of the AEF treat artillery usage with little more detail, and sometimes even less, than the air and armored forces that made their well-promoted, but clearly less operationally significant, combat debuts. This work seeks to fill this glaring gap, so that the comparison between the military efforts and experiences of the major powers of World War I can be more complete.

The historiography of the AEF can be broken down into two fairly well-divided schools of interpretation. The first camp is dominated by the AEF commander-in-chief, General John J. Pershing. He played the primary role in establishing the initial direction and tone of AEF historiography through the submission of his "Final Report" in 1919 and the publication of his two-volume memoir entitled *My Experiences in the World War* in 1931. Finally, Pershing served as chairman of the American Battle Monuments Commission (ABMC), an official government body that wrote a number of semi-official histories of the AEF. In these works, Pershing asserted that the AEF fought through numerous challenges to become an effective, well-oiled fighting machine, and that its battlefield exploits were a decisive factor in bringing the war to a close. A comparatively low total casualty figure, and the fact that the war ended less than six months after the first AEF soldiers engaged in offensive combat, enabled Pershing's assertions to go largely unchallenged for the next half century. In addition, popular histories written by patriotic journalists as well as numerous memoirs and histories from other high-ranking AEF officers reinforced Pershing's perspective in the years immediately following the war. Edward M. Coffman's *The War to End All Wars* and Harvey A. DeWeerd's *President Wilson Fights His War,* both written in the late 1960s, still adhered to the basic tenets of Pershing's perspective, although they occasionally added a rather subdued measure of criticism to the previously unblemished account. It was not until the mid-1980s that a truly revisionist interpretation of the AEF surfaced. Donald Smythe's *Pershing: General of the Armies* (1986), Paul F. Braim's *The Test of Battle* (1987), and David F. Trask's *The AEF and Coalition Warmaking, 1917–1918* (1993), as well as two journal articles by James W. Rainey (1983, 1992), tell a very different story.[4]

According to the revisionists, the AEF was not the efficient warmaking machine that Pershing, and those who accepted his story, had described. The revisionists focus on the significant problems that existed in the AEF and paint a picture of the AEF as a poorly trained, irregularly equipped, and inadequately led army. As for the combat effectiveness of the AEF, the revisionists describe American offensives in unfavorable terms, calling attention to cases of massed frontal assaults on enemy strong points and the high casualty rates and slow progress that resulted from such uninspired tactics.

Unfortunately, neither the Pershing school nor the revisionists offered anything but the most cursory explanations of the artillery forces, and their employment, in the AEF. Unsurprisingly, when the contribution of the AEF artillery forces is mentioned by the Pershing school, it is lauded for effectively providing much-needed support to the infantry. Conversely, the revisionists describe the artillery forces as being inexperienced, ineffective, and wasteful. However, in both cases the comments are usually offered without a great deal of supporting evidence. Braim devoted less than one summary paragraph to a critique of AEF artillery employment in his otherwise well-executed analysis of AEF fighting effectiveness in the Meuse-Argonne offensive. Without offering specifics, he generalized that "the artillery was poorly employed."[5] To be fair, none of these works, whether from the Pershing or revisionist perspective, has set out to explore AEF artillery employment in detail, but both compliments and criticisms have been offered none the less.

This work seeks to offer a more complex, but also more accurate, interpretation of the capabilities of the AEF's artillery forces, and eventually, of the AEF as a whole. A more focused investigation shows that the reality concerning AEF artillery employment and combat effectiveness is somewhere between the two extremes. The revisionists have provided some valuable service by pointing out that the AEF was a rather flawed military force, hastily mobilized, irregularly trained and equipped, and not at all as effective and decisive on the battlefield as the Pershing school asserted. Nevertheless, a more detailed investigation of the AEF's artillery forces, which were indeed a critical component of any World War I army's effectiveness, demonstrates that amidst the flaws and problems there was also considerable competence and effectiveness.[6]

In order to investigate the topic in sufficient depth, this work focuses primarily on just one artillery organization, the 2nd Field Artillery (FA) Brigade, which was the artillery component of the U.S. 2nd Division. Since a division was the largest autonomous combat organization in the AEF, its wartime development can be tracked (as opposed to a corps or army, whose composition of divisions changed through time, often very rapidly). In this light, the U.S. 2nd Division is the subject of this investigation, as it was one of the few AEF divisions to fight in a variety of battles, in multiple combat sectors, over the full span of time that the

AEF was engaged in active combat. The only other AEF division to see more active combat was the 1st Division. However, the 1st Division, with its 1st FA Brigade, makes a less appropriate subject because it is less representative of other AEF divisions. By all accounts, the 1st Division was treated much differently than every other AEF division that saw combat. It was the first division sent to France, arriving a full five months before most of the 2nd Division set foot on French soil. Consequently, a higher proportion of veteran soldiers were attached to it, and those soldiers received the longest and most comprehensive training of any AEF division. After the war, Major General Hanson Ely, an infantry brigade commander with the 1st Division (and subsequently with the 2nd Division), asserted that the 1st Division had "very excellent training" and that when the time for combat came it was "really ready for battle."[7] While some may disagree with Ely's assessment, one would be hard pressed to find another brigade commander who could have said the same about any other division in the AEF.

A detailed study of the 2nd Division's artillery force, from its creation and training through each of its five major engagements, provides an excellent opportunity to discover the specific strengths and weaknesses of the divisional field artillery forces of the AEF. Larger studies have approached the whole of the AEF, enabling authors to pick and choose particular snapshots from the twenty-nine AEF divisions that saw combat, and either focus on and accentuate the successes, as was done by the Pershing-led historians, or focus on the numerous instances of failure and inefficiency, as the revisionists have done. This study follows just one division throughout the whole war and lays bare the problems it struggled with as well as areas of competence and expertise.

This work is not meant to serve as a manual for successful World War I artillery employment, or as a simple narrative of one division's combat history. Though the book discusses and analyzes some of the artillery tactics and techniques used by the 2nd Division in the war, and chronicles the operational activity of that division for the duration of its World War I experience, it does so only because this method of investigation suggests certain conclusions that may be generally applicable to the rest of the AEF, and to the whole American experience in the war. This work also demonstrates the complex interaction between national theories of warfare, the military doctrines that are built on those theories, and the conflict that arises when those theoretical doctrines meet the challenges posed by the different doctrines of one's allies, as well as the harsh realities of the modern battlefield.

What emerges is also an analysis of the particular style of warfare that characterized American combat operations in World War I. The way the AEF fought was influenced by the traditions, attitudes, and organizational state of the U.S. Army at the time of the American declaration of war, by the theoretical ideals of the AEF's senior military leaders, by the myriad influences that worked on the American divisions throughout their training period, and finally by the experiences of those forces in combat. That, in fact, is the path that this book explores in the following chapters.

Chapter 1 examines the theories and doctrines that the U.S. Army carried into the war, compares them with the conflicting doctrines of its European allies, and describes how those two different perspectives influenced the training of American combat divisions. The senior officers in the U.S. Army had some very particular views about how the AEF should carry out offensive operations on the Western Front. However, due to the size and state of the U.S. Army at the time the United States declared war on Germany, it was forced to use experienced Allied officers to rapidly train much of the AEF. Allied instructors often brought along more conservative doctrinal views that were in opposition to the views held by most of the senior officers in the AEF. Those conflicting theories contained very different ideas on the role and nature of artillery employment in battle. During its formative training period in France, the 2nd Division was exposed to the conflicting theories presented by the official AEF combat doctrine and by the doctrine of the Allies. This exposure had a significant impact on the manner in which the 2nd Division was trained and in the way it would respond to the challenges of the battlefield.

Chapters 2 and 3 describe and analyze the 2nd Division's first two sets of major combat operations: the three weeks of battle near the woods and trenches around Chateau-Thierry in June 1918 and the fighting in the open fields near Soissons in July, in the Aisne-Marne counteroffensive. These battles provide the first real clash of theoretical doctrine with actual battle, demonstrate the continued impact of Allied forces on AEF operations, and show that the 2nd Division learned some hard lessons concerning the value of artillery support in World War I warfare and the difficulties associated with providing that support.

Chapters 4, 5, and 6 take the reader from those first lessons in June and July 1918, through a period of rest, refitting, and some additional training, to the 2nd Division's final three major battles. Throughout those three battles, all occurring within the last two months of the war, the 2nd Division clearly demonstrated both its capabilities and its limitations. The operations of the division near St. Mihiel in September, Blanc Mont Ridge in October, and finally in the last phase of the Meuse-Argonne campaign in November, also suggest that a particular style of warfare was developing within the 2nd Division and in other units as well. The 2nd Division demonstrated that when it had sufficient time to make detailed attack plans and was given additional artillery support, its conservative firepower-based attack style was capable of making thoroughly successful assaults on even strong enemy positions. Though it was a style that was more appropriately matched to the actual capabilities of the American combat divisions, it was not in general agreement with the theoretical views and doctrines of the senior AEF leadership.

When the war ended on 11 November 1918, this doctrinal conflict was still not completely resolved. Although there were American divisional and corps commanders who showed that they had eventually learned the lessons that the 2nd Division had been taught in its first battles, Pershing and his staff at General Headquarters (GHQ) appear to have held on to their aggressive views to the very end. The result of this conflict, between the desires and skills of the combat divisions themselves and the American High Command, was a pattern of offensive combat that did not maximize the genuine strengths of the American forces. In this sense, the conclusions of this work may generally adhere more closely to the critical assessments made by the revisionists. However, this work also attempts to demonstrate those areas in which AEF divisions proved to be very capable, and in so doing, to offer a more complete, if less simplified, understanding of AEF combat effectiveness.

NOTES

1. Report of Major General Summerall to Deputy Chief of Staff, AEF, Subject: "Artillery," 5 January 1919, V Corps Historical File, Entry 1118-32.11, Record Group 120 (hereafter RG 120), National Archives (hereafter NA).

2. For an analysis of artillery in the British Expeditionary Forces, see Shelford Bidwell and Dominick Graham, *Fire-Power: British Army Weapons and Theories of War, 1904–1945* (London: Allen and Unwin, 1982); for an excellent study of artillery development in the German Army, see David T. Zabecki, *Steel Wind: Colonel Georg Bruchmüller and the Birth of Modern Artillery* (Westport, CT: Praeger, 1994); and for a comparison of artillery development in France and Germany, see Bruce I. Gudmundsson, *On Artillery* (Westport, CT: Praeger, 1993).

3. See Fairfax Downey, *The Sound of the Guns: The Story of American Artillery* (New York: David McKay, 1955), Frank Comparato, *Age of the Great Guns* (Harrisburg, PA: The Stackpole Co., 1965), and Boyd Dastrup's *King of Battle: A Branch History of the U.S. Army's Field Artillery* (Ft. Monroe, VA: U.S. Army Training and Doctrine Command, 1992). Dastrup's work is both the most current and the most scholarly of the general histories on American artillery, but only one chapter is devoted to AEF artillery.

4. Perhaps one of the first of the revisionist accounts of AEF combat effectiveness was contained in Allan R. Millett's biographical work, *The General: Robert L. Bullard and Officership in the United States Army, 1881–1925* (Westport, CT: Greenwood Press, 1975). However, since the book was primarily biographical, and devoted less than a quarter of its pages to Bullard's World War I experiences, its impact on AEF historiography was not as significant as it otherwise might have been; James W. Rainey "Ambivalent Warfare: The Tactical Doctrine of the

AEF in World War I," *Parameters* (September, 1983), pp. 34–46, and "The Questionable Training of the AEF in World War I," *Parameters* (Winter, 1992–93), pp. 89–103.

5. Paul F. Braim, *The Test of Battle* (Newark: University of Delaware Press, 1987), p. 153.

6. One historian, Timothy K. Nenninger, has come a bit closer to this more complex, middle-ground position than most other revisionists. In his essays, "Tactical Dysfunction in the AEF, 1917–1918," *Military Affairs* (October, 1987), pp. 177–81, and "American Military Effectiveness in the First World War," in *Military Effectiveness, Volume I: The First World War*, edited by Allan R. Millett and Williamson Murray (Boston: Allen and Unwin, 1988), pp. 116–56, Nenninger tempered his generally critical assessment of AEF combat effectiveness with comments that accounted for areas of expertise, as well as some rapid improvement during the final weeks of the war. Perhaps because this perspective was contained in a journal article and a single-chapter essay, its impact on AEF historiography has not been as substantial as it might have been if it had been presented as a monograph.

7. General Hanson Ely, "Infantry in Recent Operations," lecture on 10 March 1919, to Army Center of Artillery Studies (ACAS), in Folder 22, ACAS File, Entry 371, RG 120, NA.

1

The Creation of a Modern Army: The Building and Training of an AEF Division

> It was my opinion that the victory could not be won by the costly process of attrition, but it must be won by driving the enemy out into the open and engaging him in a war of movement.[1]
>
> General John J. Pershing
> AEF Commander-in-Chief

When the United States Congress declared war on Germany on 6 April 1917, many high-ranking government officials, President Woodrow Wilson included, did not know what part the United States would play in helping to prosecute the war. Many hoped that the United States could limit its role to providing financial and logistical assistance to the Allied nations whose military forces were already engaged with the enemy—and considering the slaughter that had already occurred on the Western Front between 1914 and April 1917, it is easy to see why many carried that hope.[2] The German, French, and British armies had begun the war expecting to fight a series of battles that would be characterized by speed and maneuver, but after the initial German advance, the Allied stand on the Marne, and the famous "race to sea," in which each side sought to outflank the other, the nature of the war changed dramatically. No army on the Western Front proved capable of waging an offensive that was powerful enough to thoroughly crack the defender's lines and make a decisive advance. This situation remained essentially unchanged for the next two years of the war.

By early 1917, the opposing forces were locked in a stalemate in which little of strategic importance was accomplished except the killing and wounding of millions of troops on both sides of the trenches. While learning and change did occur, no army proved capable of waging a decisive offensive. The lack of strategic and often even tactical success on the battlefield was not the result of a lack of effort. The nations involved in the fighting on the Western Front were doing all they could to achieve victory; the unprecedented size of the forces engaged in Belgium and France attested to that. On the Western Front alone in the spring of 1917, the Allies had 190 combat divisions available to fight approximately 154 German divisions. In raw numbers of personnel, this meant that the French, British, and Belgians were maintaining a force of some four million men, and the Germans nearly three million. These troops were supported by an array of weapons that surpassed anything the world had ever seen with respect to power and versatility—machine guns, tanks, airplanes, thousands of artillery pieces, and chemical weaponry.

The European forces were even more awe-inspiring when measured against the American forces that were available to U.S. officials in April 1917. The entire U.S. Regular Army numbered less than 135,000 men, spread out in small detachments throughout the country. No permanent combat organizations larger than individual regiments existed (no brigades or divisions)—and those regiments that did exist on paper were woefully undermanned and ill-equipped.[3] General Peyton C. March, who served as chief of staff for the U.S. Army during the last nine months of the war, stated that the American Army that existed that April "was of no practical military value as far as the fighting in France was concerned; it was scarcely enough to form a police force for the emergencies within the territorial limits of the United States."[4]

At the same time as the U.S. Congress voted to declare war on Germany, British and French forces energetically prepared for another massive offensive. The commander-in-chief of the French Army, General Robert Nivelle, the primary instigator and architect of the new offensive, boasted that his attack would break through the German lines, destroy the enemy army, and end the war. Even by the standards of previous World War I offensives, this attack was to be a massive effort, and it served as a suitable description of the scale of warfare that American forces would need to be prepared for if they were to join the fray on the Western Front.

The offensive was to take place in two locations, with sixteen British and Canadian divisions attacking first near the towns of Vimy and Arras, and fifty-three French divisions advancing a week later on a front along the Aisne River (see Map 1.1). For the northern assault the British and Canadian forces were supported by 2,817 artillery pieces, and the preliminary artillery preparation lasted for five days. Nivelle's 1.2 million-man force, attacking on a sixty-kilometer front between Soissons and Reims,

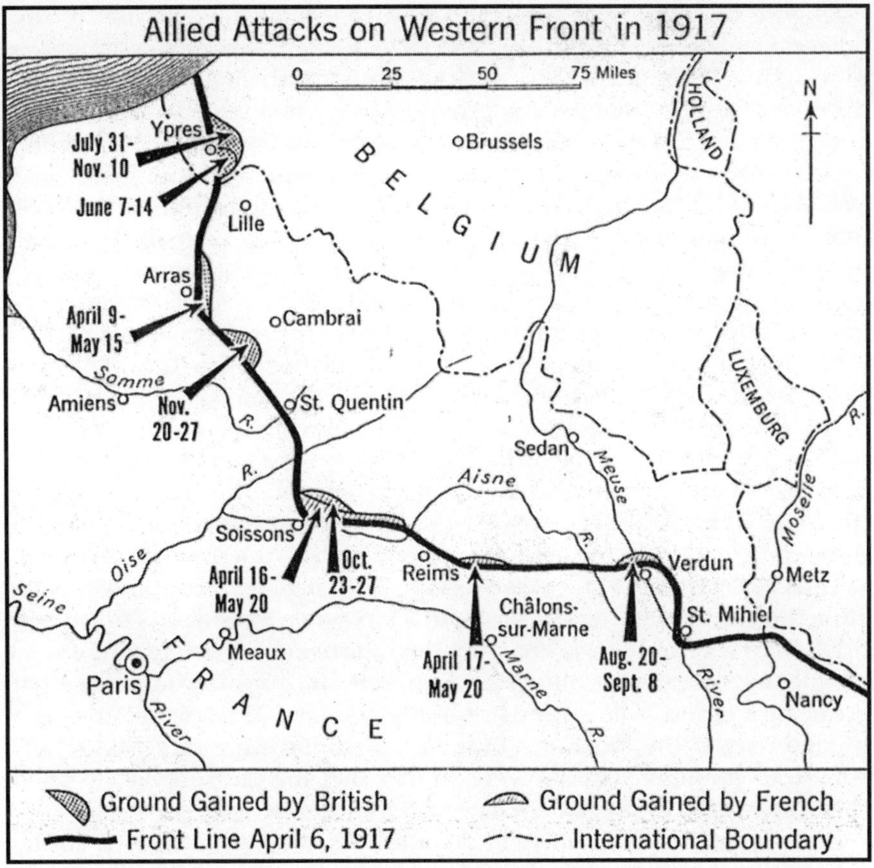

Map 1.1. Significant Allied attacks of 1917. While some attacks were more successful than others, the end result, at least strategically, was the same. The line of the Western Front hardly moved. Taken from American Battle Monuments Commission (ABMC), *American Armies and Battlefields in Europe*, p. 21.

was supported by some 7,000 guns that began their preparatory bombardment a full week before the infantry assault. These massive artillery contingents had proven absolutely necessary if the attacking infantry were to make it past the first enemy trench line without being mowed down by machine guns and pounded by the enemy artillery's counter-barrage. An American observer in France at this time wrote that "the artillery has now reached such a position of importance that successful attack or defense is impossible without it. . . . Infantry officers do not hesitate to say that infantry should not leave its trenches until the artillery preparation has really smashed all targets . . . also, the infantry can advance only so far as their artillery can escort them with fire."[5]

Despite the huge number of troops and the massive amount of firepower available for the Nivelle Offensive, neither attack even came close to breaking through the German lines with sufficient strength to gain a decisive victory. Though some ground was gained in certain areas, both attacks stalled as the troops reached the limits of the supporting artillery fire. By the end of April the British force had suffered more than 80,000 casualties. French losses exceeded 120,000, and within a matter of weeks more than half of the divisions in the French Army were in some state of "collective indiscipline," a French euphemism for mutiny.[6]

Although the field artillery on the Western Front had become an ever-increasing percentage of any attacking force, the artillery branch of the U.S. Army was just a small fraction of the relatively tiny total American force. In April 1917 there were only 408 officers and 8,253 enlisted men in the field artillery branch. In terms of leadership the branch was even weaker, as only 275 of the officers had been in the service for more than one year. There were just nine artillery regiments in the whole army (three of which had been created less than a year earlier), and none of these units had been in action since the Spanish-American War in 1898.[7]

The closest that the U.S. Army and its field artillerymen came to battle after the turn of the century was in 1916 when President Wilson sent Brigadier General John J. Pershing in command of a hodgepodge force of infantry, cavalry, artillery, and aircraft into the Mexican desert to break up Pancho Villa's bands of border raiders. The few batteries sent to take part in the Mexican Punitive Expedition were rarely used, and American munitions stocks were so low that the gunners had to shoot shells sparingly to maintain a safe supply for emergencies.[8]

Professionally, the Punitive Expedition was a veritable disaster for the U.S. field artillery branch. In order to mobilize the artillery forces required to protect the American border, the War Department closed the branch's one and only technical training center, the School of Fire for Field Artillery at Fort Sill, Oklahoma. At the same time, and for the same reasons, the U.S. Army dissolved the Field Artillery Board, which had served as the artillery branch's authority for interpreting its Drill and Service Regulations and had been responsible for monitoring and analyzing artillery information received from overseas. In March 1917, Lieutenant Colonel Dwight E. Aultman asserted in an editorial in the branch's official magazine, the *Field Artillery Journal*, that since 1916 the U.S. field artillery branch had been "professionally dormant" and that "if things drag on as they have during the past year, the professional future of the arm is dark." Just one month before his nation was to declare war on Germany, Aultman looked at the state of the U.S. field artillery force and warned that "we talk of preparedness and remain unprepared."[9]

Without any real combat experience to draw from, the only place that the artillerymen could hope to increase proficiency and improve their tactics and techniques was on the training grounds. Unfortunately, for financial and logistical reasons, artillery training was done mostly in an academic fashion. Colonel Conrad Lanza, a field artilleryman who served with distinction in the American First Army during World War I, described prewar training by saying that "except for rare experimental fires, fire for effect [meaning to actually hit and destroy a target] was not allowed; most officers had never seen such fire. . . . Consequently fire at night, or during periods of non-visibility, or against concealed targets . . . was prohibited. Such fire was considered a waste of ammunition." While this was in large part due, as Lanza said, "to the small amount of artillery . . . and the smaller amount of ammunition provided for training," it was also caused by a U.S. Army doctrine that consistently minimized the importance of artillery to combat operations.[10]

Army doctrine, as expounded in the Drill and Service Regulations, was built on the previous experiences of the U.S. Army, most notably in the American Civil War and numerous Indian wars. It stressed the use of maneuver in aggressive offensives, in which American forces were to break the enemy formation and then rout them with a determined pursuit. With this nineteenth-century view of warfare, the regulations could say with confidence that the artillery "has in general *no independent role* on the battlefield. Its *sole object is to assist* the other arms [i.e., infantry and cavalry]. To secure *decisive* results, troops must advance, occupy the hostile position and by *vigorous pursuit* destroy or throw into confusion the hostile forces. *The isolated and independent action of Artillery leads to no decisive results* (emphasis added)." The regulations stressed over and over the need for field artillery officers to use "good judgment and great versatility" so as to provide support "without undue expenditure of ammunition."[11] It is significant that while in Europe the artillery forces were expending millions of shells in support of single offensives, U.S. Army regulations treated the economic use of ammunition as among the most important concepts to be mastered by American gunners.[12]

Despite its anachronistic flavor, American doctrine was not produced in an environment devoid of information about the war in Europe. The U.S. Army had a number of military advisors and attachés monitoring the fighting in Europe, and all reported how drastically modern warfare had changed, especially concerning the role of artillery. At the same time that the 1917 edition of the Drill and Service Regulations was being published and disseminated, the *Field Artillery Journal* printed several articles that described the position of importance that the artillery forces had achieved in war. One lengthy article, entitled simply "Field Artillery Notes," described in great detail the size and scope of World War I artillery operations and brought to the U.S. Army a clear exposition of the reigning artillery doctrine in Europe in 1917, the doctrine of "destruction."

> No attack is possible until after an intense and effective artillery preparation, which has for its objects: (a) to *destroy* the enemy's barbed wire; (b) to *disintegrate* and *destroy* the enemy's trenches and dugouts, and to *destroy* or *annihilate* their defenders; (c) to prevent, or at least to interfere with, hostile artillery action; (d) to prevent the passage of the enemy's reserves by curtain (barrage) fire; and (e) to *destroy* the machine guns where ever they can be located. (emphasis added)[13]

This doctrine was decidedly at odds with contemporary American artillery doctrine, which stressed the use of light gun batteries to provide timely but occasional close-range support to small, mobile infantry units.[14] This discrepancy was quite clear to some field artillery officers back in the United States. For example, Colonel Aultman exclaimed that "new methods" of artillery employment had evolved and that "in every sense the doctrine of Field Artillery has advanced." He asserted that "the progress of the arm has been unprecedented, so that the organization and methods of three years ago are now in many respects obsolete." After comparing the state of American artillery organization and official doctrine to reports from Europe, Aultman felt compelled to admit that "during this time the U.S. Field Artillery has stood still."[15] He might have added that the rest of the U.S. Army was also only keeping pace with its field artillery branch. In short, when America joined the war, the whole of its army, and especially its field artillery branch, was too small, devoid of any applicable combat experience, insufficiently trained, and in possession of a doctrine that did not appear to be in any way suited to the daunting military operations that lay in its future in Europe.

The man given the responsibility for correcting these deficiencies was Major General Pershing (see Figure 1.1), who, along with a small staff, traveled to Europe in late May 1917. He carried with him a letter from Secretary of War Newton D. Baker, which designated him as the commander "of all the land forces of the United States" that would operate in Europe.[16] After viewing the challenges of the Western Front, it did not take Pershing long to assess the size of the force that the US would have to provide to be of any real operational assistance. By the first week of July he had requested that the War Department plan on sending him "at least 1,000,000 men" within the next year. A week later, Pershing increased his assessment by warning that Washington should base future plans "on three times this force, i.e., at least three million men."[17] Pershing and his staff eventually decided to organize the AEF into divisions that were 28,000 men strong. Each would contain over 13,000 infantry-

Figure 1.1. General John J. Pershing, commander-in-chief of the American Expeditionary Forces. Courtesy McCormick Research Center.

men in its two infantry brigades (each brigade with two regiments), and 72 artillery pieces in its artillery brigade. The 5,055-man artillery brigade was to be divided into three twenty-four gun regiments, two armed with light field guns of the "three inch" or 75mm variety, and one with "six inch" or 155mm howitzers.[18]

The AEF combat division had some significant features, most notably its total size; at 28,000 men, it was at least twice the size of most fully manned French and British divisions. One reason for creating such a large division was that it allowed the AEF to get by with a minimum number of senior officers in command and staff positions. Certainly, the tiny peacetime force had few to offer in the short term. However, Major General James G. Harbord, Pershing's chief of staff and a future commander of the 2nd Division, stated that when Pershing and his staff were searching for the proper size for their divisions, the principal consideration "was tactical." They believed that in order to effectively fight in the trench warfare environment, in which it was "quite a problem" to replace divisions that were in line, AEF divisions had to be capable of continuing to fight for "some time." They were convinced that "no *decisive stroke* could be secured" without a division that was strong enough to make penetrating attacks for "several days." That way it could continue to advance, without interrupting "the continuity of the battle . . . before *a decision was reached*" (emphasis added).[19]

Here is the first hint of Pershing's views on how the AEF was to fight in Europe, and it was somewhat of a mixed signal. Pershing demonstrated an awareness of trench warfare conditions, particularly the heavy casualties that would be sustained and the difficult logistical problems associated with replacing divisions already in line. He therefore created a division that in the words of one historian was "a behemoth" built "for sustained slugging."[20] However, this huge division was not just to grind forward in a battle of attrition, it was to "penetrate" the enemy lines and provide a "decisive stroke." Pershing had seen what trench warfare was like and had made a direct adjustment to his organization, but he still appeared to be holding on to the prewar army doctrine that focused on breaking the enemy formation and aggressively pursuing the retreating enemy forces until they were destroyed.

Another sign that Pershing and his hand-picked staff might not have fully grasped the nature of warfare on the Western Front is found in the reports of an early meeting between the AEF staff and a special War Department fact-finding team called the Baker Mission. After visiting numerous French and British organizations, the Baker Mission met with Pershing and some of his staff to discuss its findings before making a report to the War Department. The two teams were in general agreement on nearly all matters—except the size and strength of artillery support for attacking divisions, which would prove to be another important feature of the large AEF combat division.

The senior artillery officer in the Baker Mission was Colonel Charles P. Summerall, a future brigade, division, and corps commander in the AEF, who would become chief of staff of the U.S. Army after the war. Convinced that a lack of firepower in Pershing's large divisions would cost men's lives, Summerall vehemently urged that the AEF plan to provide its divisions with twice the number of guns allocated in the AEF staff plan (see Figure 1.2).[21] Summerall's artillery estimates called for each attacking division to be supported by 259 dedicated guns from the divisional, corps, and army artillery groups. The AEF staff came up with a figure of just 122 guns for the same scenario. General Harbord reported that Pershing, and the artillery officers on his staff, considered Summerall's recommendation to be "a disproportionately large amount of artillery."[22] Considering the manner in which Pershing envisioned using his artillery, his position in this matter is not surprising.

Although Pershing's staff ensured that the Baker Mission carried its

Figure 1.2. A typical AEF field artillery brigade with most of its manpower and weapons. Six of the brigade's twenty-four 155mm howitzers are up front and the forty-eight 75mm guns are on the sides. Not included in the photo are the thousands of horses used to haul the guns and ammunition to and through the battlefield. Courtesy McCormick Research Center.

recommendations concerning artillery support back to Washington, Summerall's opinion was clearly expressed in the Baker Mission's final report.

> It, therefore, becomes a question as to whether victory should be obtained through a long and exhaustive period, after great losses in men and money, or whether success should be sought through the use of a maximum of gun fire over a shorter period and with a minimum cost in lives. *It may fairly be stated that losses in war today are inversely proportional to the volume and the efficiency of friendly artillery fire.* If we are to produce a decided effect upon the issue of the war, we must strive to develop some form of rolling offensive over a very considerable area and for this purpose, artillery must be furnished in quantities not hitherto contemplated. (emphasis added)[23]

If Summerall argued his case to Pershing in words similar to those in the mission's report, it is no surprise that he was unable to convince the commander-in-chief to accept his scheme for artillery organization. Summerall's plan for artillery support envisioned, if anything, an even greater dependency on massed firepower than that exhibited by the Allied armies in 1917. While Summerall might have been unaware of it at the time, his ideas ran directly counter to the "open warfare" doctrine to which Pershing was becoming increasingly attached.[24]

With the organizational decisions made, Pershing had to focus on securing sufficient material to equip his embryonic army. Since there was to be a considerable delay before factories in the United States would be able to manufacture their own guns in sufficient quantities, Pershing secured an agreement from the French to provide all the artillery pieces the AEF required until American industry could meet its needs.[25] This meant that the artillery regiments in AEF divisions would be armed with the famous rapid-firing French 75mm gun (see Figure 1.3) and the effective, but less mobile, 155mm Schneider howitzer.

Both French weapons were acceptable field pieces, but each had limitations. The French 75mm was an excellent, reliable gun that could fire more than twenty aimed shells a minute. It was also light enough to be moved into position by just a few horses, or even a few men if the distance was not too great. It fired a rather light 13 lb shell, which, while a logistical advantage, meant that its explosive power was not very impressive, especially against well-entrenched defenders. The 155mm Schneider was at the opposite end of the field artillery spectrum. While the 75 was a very light piece with limited power, the 155mm Schneider had no trouble destroying enemy trenches with its 130 lb shell. However, the price of that great power was paid in weight—the 155mm required an eight-horse team just to move the weapon itself, even over good terrain. Transporting the heavy shells was an additional logistical challenge.

Figure 1.3. AEF gunners firing one of their famous French 75mm guns. Courtesy McCormick Research Center.

While AEF forces would enjoy the strengths of the French weapons, American infantry would also occasionally suffer from the light power of the 75mm and from the poor mobility of the 155mm (see Figure 1.4).[26]

The decision to use these French weapons, while sufficient to meet the AEF's material needs, had other important consequences. First, since it was sure to take a long time before suitable quantities of French guns could be installed in all the training camps in the United States, American artillerymen would have to be retrained on the new guns when they arrived in France. Second, and perhaps most important, the training of AEF field artillery units was certain to be dominated by French officers, and greatly influenced by French doctrine.[27] With very few available options, AEF Headquarters directed that the first artillery units brought to France (the artillery regiments for the 1st, 26th, 42nd, and 2nd Divisions) begin training at French artillery camps, with French materiel, and under the tutelage of French instructors. Pershing's staff also advised those first regiments to use certain French signals and methods of communication. Thus began an intimate relationship between the field artillery branches of the AEF and the French Army, a relationship that in the eyes of many senior AEF officials was a mixed blessing.[28]

Figure 1.4. The French-built 155mm Schneider howitzer served as the heavy artillery support for each AEF division. While it provided excellent fire-support in set-piece attacks, it was extraordinarily difficult to move through the muddy moonscape of No Man's Land, even with a team of eight horses. Courtesy McCormick Research Center.

Immediately following the U.S. declaration of war, the War Department, and later the AEF, began a furious search for information on Western Front warfare. Thanks to the British and French military missions that quickly set up shop in Washington, the War Department was flooded with regulations, pamphlets, and field notes that described in great detail the Allied approach to every facet of trench warfare. It comes as no surprise that much of this information dealt either specifically or generally with the use of artillery. The understaffed teams of American officers at the War Department and the Army War College received, translated, edited, and immediately published much of this information, apparently not thinking that it might prove contradictory to the approved doctrine of the U.S. Army, especially as interpreted for the AEF by General Pershing.[29]

One such pamphlet, "Instructions for the Training of Divisions for Offensive Action," printed practically unedited from a British document, clearly demonstrated the conservative Allied view of the infantry's absolute reliance on artillery support. It stated that "the furthest objective allotted to any troops in the Division must be within range of effective Artillery support." The artillery was expected to "*demolish* the enemy's trenches . . . *destroy* wire entanglements and other obstacles . . . cause casualties and loss of morale among the [enemy] garrison . . . *destroy* or block communication trenches . . . and to *destroy* the enemy's Artillery" (emphasis added). While playing up the role of the artillery, the document stressed the weakness of the infantry. Leading infantry battalions were never to be expected to take objectives that were more than 1,200 yards to the front and they were to always immediately "consolidate" their gains so as "to make certain of every piece of ground won." Finally, commanders were cautioned "not to ask the troops to do too much."[30] These conservative concepts were in direct opposition to the principles that Pershing would espouse.

The French military mission in the United States was especially zealous in sending artillery officers to the artillery training camps in America. When Major General William J. Snow was appointed to the post of chief of field artillery for the U.S. Army in February 1918, he found that there were between seventy and eighty French artillery officers in America, "one or more at every field artillery camp, in addition to those on duty with the [French] Mission in Washington." Speaking of these liaison officers, Snow asserted that they often "had been but a little time in the service, were not well grounded in field artillery, had left France at the height of stabilized warfare, and consequently knew no field artillery except the trench phase."[31]

Snow and many other senior artillery officers in the United States were convinced that the trench-based warfare that had characterized the fighting on the Western Front since the end of 1914 was but an extended anomaly that was sure to end before the war did. Colonel A. S. Fleming, the commandant of the reopened School of Fire, told incoming students that "none can say when trench fighting will give place to open warfare, and *when* it does, the army that is not skilled in the latter is a beaten army." While the urgent need to get even "partially educated" artillery officers to France was too great to allow training camps in the United States to suitably teach the important details of tactics and interarm liaison, instructors in American artillery camps did strive to teach "the principles of *open firing* and their application to the particular cases arising in trench warfare" (emphasis added).[32] Senior artillery officers in the United States were convinced that they needed to teach more than just the basic techniques of shooting and certainly more than only those trench warfare methods that the French officers focused on. However, due to the urgency of the situation, most artillerymen were sent overseas having only learned how to shoot the American version of the French gun that they would use in France. Tactics, liaison, and mobile operations had to be learned overseas.

In France, Pershing's staff was hard at work developing the training regimen for the units that would join the AEF. Eventually, AEF Headquarters decided that the artillery regiments would move through three distinct phases of training. The first month would focus on "preliminary instruction," to include familiarization with and the firing of the new French weapons, the training of specialists in the use of wireless, telephone, and repair of materiel, and the use and care of animals. The second month was to consist of a tour at a "quiet" sector of the front lines under the care and tutelage of a French unit. There the officers would learn to care for their troops, and the batteries could gain experience firing at real enemy targets in a combat environment. After completing its tour at the front, the whole division was to assemble together for a month of maneuvers, during which the artillery was to work on issues of liaison and communication, as well as review important material from the first two phases. During this period the American divisions were to hone their ability to engage in what Pershing and other American officers called "open warfare."[33]

While his staff was developing the training network and procedures, Pershing began to elucidate the combat doctrine that was supposed to guide both the training and the fighting of his divisions. The more Pershing saw of trench warfare, the more he hated it. He arrived in France in time to experience the French depression and mutinies that followed the costly failure of the Nivelle Offensive, and then witnessed the even more costly floundering of the British forces in Field Marshal Sir Douglas Haig's four-month-long Passchendaele Offensive in Flanders. In the latter effort, the Allies advanced their lines just eight kilometers at the cost of approximately 300,000 casualties. Seeing the worst of what trench warfare had to offer, Pershing wanted none of it, and he became increasingly convinced that he and his American "doughboys" had to employ some other method to fight the Germans.[34]

Pershing was clearly unimpressed with the French and British offensive doctrines, which he said were "based upon the cautious advance of infantry with prescribed objectives, where obstacles had been destroyed and resistance largely broken by artillery." He thought the Allies were overreliant on artillery support and that they did not place enough emphasis on getting the most from their own infantry. Pershing criticized the French infantryman for not relying on "his rifle" and for not making use of "its great power." After a relatively short time in Europe, Pershing became convinced that the long periods of trench fighting not only were extremely costly in men and material, but that they also had a terrible "resultant psychological effect" on Allied troops. He believed that trench warfare sapped the aggressiveness of an army and had led the Allies to accept the conservative, artillery-centered doctrine that he thought guided both their training and their ineffective offensives.[35]

If there was one characteristic that Pershing was determined to instill into the AEF, it was aggressiveness, both in the strategic and the tactical sense. Plans for the AEF's major operations were to be based on the premise that in order "to bring about a decision the [enemy] army must be driven from the trenches and the fighting carried out into the open." Just how his forces were to drive the enemy out of the complex trench systems, often many kilometers deep, was not said. Pershing seemed to assume that after a successful assault on the enemy front lines, the enemy would be compelled to commence fighting "in the open" rather than reassuming a defensive posture behind new positions.[36]

In making this assertion Pershing stood squarely on the principles explained in the Army's current Field Service Regulations, which assumed that an open "pursuit" phase immediately followed any successful assault. The regulations stated that, "Only by means of an *energetic pursuit* of the beaten enemy can the full fruits of victory and *decisive results* therefore be obtained. . . . The task of the victorious army is less than half performed when it remains satisfied with the mere possession of the battlefield." In the post assault phase, "*boldness to the point of rashness* is not only justified but is demanded" (emphasis added).[37]

The service regulations clearly stated that "aggressiveness wins battles," and Pershing was intent on proving it. Pershing demanded that all AEF plans have as their main object the assumption of "a vigorous offensive," in which the AEF would replace the conservative trench warfare methods with the aggressive American combat doctrine of "open warfare." Unfortunately, neither Pershing nor his staff ever clearly and definitively elucidated what exactly "open warfare" meant. Almost all senior American officers spoke of it, but just how it was to be inaugurated on the Western Front was never made clear. In fact, as late as April 1918, Major General Hunter Liggett, an American corps commander, wrote to GHQ and informed them that he could find "nothing in the mass of literature" he had been given that explained exactly how his division commanders were to prepare their units for "open warfare."[38]

Even Major General Harbord, who served as Pershing's chief of staff for a time, could only offer this poor explanation of the difference between trench and open warfare:

> [T]he War would never be won by troops of both sides remaining in parallel trenches separated by a few hundred yards. Some day someone somewhere would come out of his trenches and start forward and thus a stalemate would be broken and the War would eventually be won. When even one soldier climbed out and moved to the front, the adventure for him became *open warfare*, a war of movement, and the essentials of minor tactics were then in play. (emphasis added)[39]

Of course, with such a nonsensical description of "open warfare," one could assert that "open warfare" had continued unceasingly from the beginning of the war right through to 1918. Later, Pershing wrote that the absence of the rolling barrage was one of the defining characteristics of "open warfare," at least as far as troop formations were concerned. Others in the AEF held different views on the use of artillery in "open warfare." A training memorandum in the 2nd Division stated that the rolling barrage would still be used in "open warfare," but that, instead of being based on "a fixed timetable," it had to be arranged by signals during the fighting, such as colored flares. There were to be but two signals for the use of the rolling barrage in open warfare: "one to halt the barrage, another to lift the barrage."[40] Needless to say, there was a great deal of ambiguity concerning open warfare, especially what it actually was, how artillery was to be used while engaging in it, and how it was to be forced on an enemy that was determined to use one strong defensive position after another.

In fact, the other senior Allied commanders also hated trench warfare, and longed to be done with it. Both Nivelle's 1917 offensive and Haig's Passchendaele Offensive were attempts, at least initially, to break through the enemy's trench system and engage enemy forces in a war or movement out in the open.[41] One could go so far as to say that with the exception of the German attack at Verdun in 1916, every major offensive on the Western Front—from late 1914 to the British attack of Messines Ridge in mid-1917—had as its strategic purpose the driving out of the enemy into the open. The problem during these years was not that Allied commanders desired to fight in a trench-to-trench style, but that for a number of reasons, they lacked the capability to break the stalemate. The conservative artillery-based attack doctrines that had developed in both the British and French armies was not so much a repudiation of the "open warfare" style as a response to the years of failure associated with attempts to bring it about.

Although there were questions concerning the AEF concept of "open warfare," Pershing did make it clear just which branch of the army was the decisive factor in its execution, and that was the infantry. The Allied slogan "artillery conquers, infantry occupies," was to have no place in the AEF. Pershing insisted that the AEF look forward to "an *aggressive offensive* based on *self-reliant infantry*," (emphasis added) and he constantly urged commanders to train their divisions accordingly. Commanders were to ensure that their infantry were so filled with an "aggressive spirit" that each "soldier feels himself, as a bayonet fighter, invincible in battle."[42]

Here again, Pershing did little more than apply the main tenets of the army's existing regulations, which stated that "the infantry is the principal and most important arm, which is charged with the main work on the field of battle and decides the final issue of combat." Convinced that "the general principles governing combat" remained "unchanged in their essence," Pershing insisted on applying the army's prewar doctrine with as few alterations as possible.[43]

Pershing sent a flurry of messages back to the War Department passionately instructing it to base all training in the United States on the army's infantry-focused "open warfare" doctrine.

> Close adherence is urged to the central idea that the essential principles of war have not changed, that the rifle and the bayonet remain the supreme weapons of the Infantry soldier and that the ultimate success of the army depends upon their proper use in open warfare. To this end it is recommended that intensive training in all phases of open warfare be accepted as the principle mission of divisions before embarkation.[44]

As for the artillery, Pershing similarly asserted that "the fundamental principles governing conduct of fire for Field Artillery remain essentially unchanged." Therefore, while the artillery forces of the Allies were focused on perfecting standing, rolling, and box barrages, with a corresponding emphasis on accurate map firing, Pershing advised his training coordinators to focus on different activities. He stated that the "most important" artillery skills for open warfare were "mounted instruction over difficult ground and at increased gaits, care of horses, . . . rapid preparation and conduct of fire *without maps*, rapid reconnaissance and occupation of positions under tactical situations, and changes of position by echelon" (emphasis added).[45] Pershing envisioned his forces fighting in a very different style from that employed by the Allies, especially concerning the employment of artillery in the attack.

By late 1917 both the British and the Germans were beginning to develop styles of attack that lay somewhere between total reliance on weeklong, crushing artillery barrages and Pershing's open warfare concept based on infantrymen fighting their way forward with rifles and bayonets. The Germans were developing a system based on a combination of highly trained, well-armed infantry and massive artillery support that could drive into and through enemy positions. The British were also slowly developing a combined arms team of infantry, artillery, tanks, and even attack aircraft that could successfully drive forward and take even the most heavily defended enemy positions. Because this British style was not codified as doctrine during the war, and was only fully implemented in the final six months of the war, it was not readily accessible to Pershing. Nevertheless, it is likely that Pershing would not have appreciated the heavy reliance on artillery the British system called for, or the rather set-piece battles that it successfully waged. It certainly was not a system that could "drive the enemy out in the open" or "engage him in a war of movement."

There is some evidence that Pershing and the other "open warfare" proponents in the U.S. Army were aware that strong artillery support was necessary to effect the initial breach in the enemy lines of resistance, through which the American infantry would advance and begin a vigorous exploitation. The open warfare doctrine appears to have been based on the theory that the infantry would be essentially "self-reliant" after that point, and that what little support they did need from the artillery could be provided by light gun elements that galloped into position, set up in the open, and fired off a rapid volley to smash any troublesome pocket of resistance. If infantry losses were slightly higher due to a reduction in available firepower from artillery, this was thought to be more than made up for by not allowing any substantial stabilization to occur.[46] However, before the AEF actually engaged in combat operations, most of its senior officers put little emphasis on the importance of massive artillery support, even in the initial phase of the assault. Perhaps it was just taken for granted, but more likely the absence of any meaningful discussion about the crucial role of artillery was the result of a doctrine that accentuated the capabilities of the aggressive, "self-reliant" infantryman and underestimated the importance of heavy firepower.

This doctrine, and the style of warfare that it encouraged, had all but disappeared from the Western Front. The size and scope of the modern battlefield, coupled with the lethality of the defending force's firepower, had made that style of warfare practically impossible to effectively carry out. In order to successfully engage in the flexible, maneuver-based warfare that Pershing so desired, the combat units needed to be able to break through the formidable defenses and demonstrate an extremely high level of expertise in the areas of combat liaison, mobile communications, transportation, supply, and the rapid but coordinated employment of firepower. While it had been a challenge to effectively carry out these functions in previous wars, as of 1917, the technological development in these fields had lagged behind that of the killing power of modern weaponry, making the proposed American style of warfare extremely difficult, if not practically impossible, to put into practice.

Having formulated the combat doctrine of the AEF, however vague in parts, Pershing's next challenge was to train his forces to carry it out on the field of battle. With the army undergoing rapid expansion, and with the Allies screaming for the Americans to get into action, the task was certain to be even more difficult. Pershing himself admitted that "the most important question that confronted us in the preparation of our forces of citizen soldiery for efficient service was training."[47] However, it should be noted that numerous American officers, especially artillery officers, stressed that it was much more difficult to train forces for open warfare than for trench fighting. An article written at the reopened School of Fire asserted that "any officer who has mastered open warfare . . . could get in a hole in the ground and take care of the situation at hand. But an officer who is master of only the comparatively slow and very accurate methods of position warfare would be lost in open warfare."[48] The reasoning was that in trench warfare a small number of skills were used over and over, almost always with plenty of time to prepare, and with very specific guidance from commanders. However, in open warfare artillerymen were required to use a myriad of skills, to execute numerous missions rapidly without any opportunity to plan them in advance, and regularly do so without the constant and specific involvement of senior officers. Therefore, training artillerymen to perform well in the less structured environment of open warfare took a good deal more time than to teach them the important details of trench gunnery.

Pershing wrote that open warfare was based on "individual and group initiative, resourcefulness, and tactical judgment," skills that were not easily or quickly developed, especially in raw recruits.[49] It thus appears that the American doctrine demanded a great deal of its brand-new, untested training system, quite possibly more than the system could provide considering the material it had to work with and the short time in which it had to produce the results. Certainly, everything would have to go almost perfectly, and it did not.

The first challenge to be overcome was the weak initial training level of the American men and units sent to France. The three regiments of the 2nd Field Artillery Brigade were all created after the declaration of war and were formed by a combination of raw recruits and a small, experienced cadre taken from one of the few regular prewar regiments. When the 12th, 15th, and 17th Field Artillery Regiments of the 2nd Brigade gathered at the French artillery camp at Valdahon in January 1918, the percentages of troops who were in their first year of military service were 84, 87, and 90 percent, respectively. Their only training, besides the basic military training they had received when entering the service, was comprised of standing drill and the occasional firing of their American weapons. Thus, it is safe to say that they were not proficient in either trench firing or open warfare operations when they disembarked in France.[50] Some of the junior officers destined to become battery commanders, like Lieutenant John D. Clark, had been sent to France to serve in other branches, such as the ambulance service, and then were absorbed into the field artillery arm and sent to Valdahon without any previous artillery experience.[51] The low overall level of experience and aptitude was important for a second reason. Without any knowledge or commitment to a previous American style of fighting, many American troops were blank slates, ready to be taught and molded by the French instructors.

After arriving at the Valdahon camp, the artillerymen were immediately turned over to a cadre of French instructors who introduced them to their new weapons, the French 75mm gun and 155mm howitzer, and taught them "the French methods of firing."[52] Batteries fired on the range from 7:00 AM to noon, six mornings a week, and spent the afternoons taking in theoretical instruction, running through drills, and working in the ordnance shops. Unlike many senior officers back in the United States, American artillerymen at Valdahon seemed to be rather impressed with their French instructors. Brigadier General George L. Irwin, who commanded the 2nd Field Artillery Brigade during its training period in France, described the French instructors as "officers of experience, devoted to their profession, and eager to render every assistance to their allies." Irwin did note, however, that the "long period of stationary trench warfare" had left the French instructors with "a very palpable disregard of the methods necessary in a war of movement." Statements such as this suggest that few open warfare concepts were taught to or practiced by the 2nd Brigade while at Valdahon.[53]

All other indications agree that this first period of training, which lasted for nearly two months for the 15th and 17th FA Regiments and about six weeks for the 12th FA Regiment, focused on techniques that pertained more to trench warfare than to open warfare. According to the final report of General Ernest Hinds, the AEF's chief of artillery, the focus of this training was on the "new," or more scientific "methods of fire," involving "the calculation of firing data from the map, meteorological and ballistic corrections, precision adjustment, etc." Those were hardly the types of skills that Pershing listed as being critical to artillery operations in open warfare, but they were the techniques that the Allied artillerymen were employing in their more conservative, artillery-dominated attacks.[54]

Though this was anathema to Pershing, it seems that the AEF could hardly have done otherwise. First, the untrained American artillerymen definitely needed to learn the basics of the modern methods that were used to provide accurate fire in both defensive and offensive warfare on the Western Front. Second, the dearth of trained American artillerymen precluded any effort to use American officers to train these first units on their own. General Hinds stated that the AEF was "tremendously handicapped" by a lack of experienced American instructors. This meant that the first AEF artillery brigades would be trained almost exclusively by French officers. This was in fact the case with the 2nd Brigade while at Valdahon, where the staff of instructors was "entirely French" until 1 April 1918, when one American officer finally arrived to act as director of the camp. The results were therefore predictable; as General Irwin alluded to, the French officers taught those skills, techniques, and methods of artillery employment that they knew best and were convinced were most necessary.[55] Obviously, this did very little to prepare the American artillery units to fight in the open warfare style prescribed by official AEF doctrine.

One other factor undoubtedly forced the training of the 2nd Brigade to focus on trench, rather than open, warfare methods. A review of the skills that Pershing listed as being most crucial to the effective employment of artillery in open warfare clearly shows how important the use of horses was to properly train field artillery units. The problem was that many artillery brigades, the 2nd Brigade included, did not receive their allotment of horses until just before leaving for the front to begin the second phase of training. It was therefore impossible for them to work on "mounted instruction," "rapid reconnaissance and occupation of positions," and echeloned position changes. Open warfare was based on mobility, and without horses the batteries of the 2nd Brigade were quite immobile. This would have reinforced the tendency for the brigade to focus almost exclusively on trench gunnery methods.[56]

While unable to train for open warfare, the 2nd FA Brigade did receive quality instruction in techniques necessary to provide support for conservative, artillery-centered attacks more commonly associated by American officers with the attritional strategy of the Allies. The three regiments completed their first phase of training by accomplishing a brigade-level exercise in trench warfare methods. As part of a "simulation game," the brigade fired typical defensive barrages, shifting the fire "according to a pre-arranged scheme." The exercise was carried out completely "under trench warfare assumptions."[57] The 2nd Brigade appears to have learned the trench methods well, and it left the camp for its second phase of training with its commander claiming that the brigade's progress was "eminently satisfactory."[58]

The 2nd Brigade accomplished its second phase of training at a "quiet" sector of the front just south of Verdun. During the last week of March 1918, each regiment was assigned to the area of a different French division and placed under the control of the appropriate French divisional commander. American batteries were immediately assigned to work with experienced French batteries already in the sector. This arrangement further solidified the trench warfare methods that had been taught at Valdahon. In order to improve the skills of American officers assigned to the brigade and regimental staffs, they were allowed to visit the headquarters and command posts of the French divisions and corps under whom they served.[59]

While the training plan directed that American commanders and units slowly be granted more responsibility at the front, the need for French divisions to help stop the first two German "spring offensives" accelerated the schedule.[60] In April, the 2nd Division, infantry and all, was given its own sectors and American batteries were expected to handle requests for protective barrages on their own. The 12th FA claimed that during this period its troops "learned a great deal about position warfare," and it appears they did. Though the division had originally been scheduled to leave the front in April so that it could spend a full month in open warfare training, the 2nd Division had to remain in its front line sector until 12 May, when it was relieved and sent to a billeting area in the rear.[61]

At this time the first two phases of training had been accomplished, but between them neither the division nor its artillery brigade had done any training in open warfare. It had not had the chance. It had spent six to eight weeks learning French artillery methods in a training camp, during which it had no horses, and then another eight weeks perfecting those methods in an environment that exemplified at least one form of trench warfare.[62]

By all indications the artillery regiments progressed well during this "trench-oriented" training. Pershing commented favorably on the division when he inspected it during its tour at the front. Perhaps of greater significance, the French corps and army commanders under whom the 2nd Division had served all rated the division very highly. Pershing later claimed that General Hirschauer, the commander of the French Second Army, told him that "without a doubt it was then as efficient as any of his French divisions."[63] However, in terms of being able to engage in open warfare, the division had made no progress. That was to be the focus of the third phase of training.

While the third phase of training was originally designed to last for a full month, the crisis caused by Germany's spring offensives and the urgency to get American divisions ready for unlimited and unsupervised front-line duty changed the schedule. By the time the 2nd Division left the quiet sector near Verdun, its third phase of training had been shortened to a six-day period of terrain exercises and unit maneuvers.[64] After arriving in the training area on 22 May, the division settled in and finally began the open warfare training that was supposed to be the foundation of the AEF style of war. The six days were spent mostly in command exercises (i.e., without troops) and regimental and brigade maneuvers. On just one of the six days was the entire division assembled and allowed to engage in full divisional maneuvers. Nevertheless, by 29 May, the division had received orders to move to Montdidier to replace the U.S. 1st Division in line near Cantigny. However, before that movement was made, new orders were received. On 27 May the Germans had launched a third massive offensive, this time in the Chemin des Dames area, and within two days their stormtroopers made a penetration some twenty-five kilometers deep and seventy kilometers wide. The new orders for the 2nd Division directed it to advance towards Chateau-Thierry to help stop the German attack. The time for training was past, it was time for the division to demonstrate what it had learned.[65]

There can be little doubt that the units of the 2nd Division did *not* receive sufficient training in Pershing's open warfare methods. Of the roughly sixteen weeks of training given to the 2nd Division in France, only one was in any way devoted to teaching the skills that would be most important in an attack based on rapid maneuver and flexible artillery support. And it should be remembered that all of the instruction received during the first two phases came from French officers, who by and large knew almost nothing of open warfare, and who cared even less. Those French officers, both in the camp at Valdahon and in the trenches near the front line, could only transmit their conservative, position-oriented, artillery-focused doctrine, because that is what they knew.

Not only was the conservative, artillery-focused doctrine of the French indirectly transmitted to American battalion and battery commanders while they were trained by French officers in artillery camps and French trenches, but it was directly taught to American staff officers in the AEF's General Staff College. In the staff school, French and British officers gave crystal-clear descriptions of how critical artillery support was to warfare on the Western Front. They taught what was to be expected of the supporting artillery forces and what could, or could not, be accomplished without them.

One lecture on the use of artillery in an offensive taught the staff officers that it was "necessary all during the attack to keep under fire: (a) the immediate objectives of the attack, (b) the flanking defenses which could paralyze the attack, and (c) the rear lines from which the enemy could fire on the attacking forces." It was also important to "provide for the protection of the infantry after the attack."[66] Another lecture stated very clearly that because of the necessity to have overwhelming artillery support for every assault, "divisions can progress only step by step, toward limited objectives." Since the defensive positions were so strong, and attacks had to have limited objectives, "preconcerted plans covering all phases of the attack, far from being objectionable, now prove to be the only ones likely to meet with any success." This last point, a clear indication of the move toward set-piece battle plans in the Allied armies, directly contradicted the American Field Service Regulations, which stated that such "preconcerted plans" were, in fact, "objectionable."[67]

Even when discussing the exploitation, or pursuit phase, which was when Pershing's infantry-focused open warfare methods were supposed to be most effective, the staff school lectures accentuated the role of artillery support and the dependence of the infantry on it. In the lecture "Exploitation of Success," staff officers were taught that while successes had to be "thoroughly exploited" to be of any real value, it was still a "necessity for infantry to be constantly supported by artillery."[68] Another lecture warned the students to "never venture Infantry in an attack against organized positions without having carefully prepared this attack. And when I say organized positions I do not only mean trenches and wire entanglements. I also mean strong points, hastily organized as they may be; every possible strong point which has to be attacked or outflanked must be taken under strong artillery fire. It is the only way for preventing tremendous and useless losses."[69]

All of these principles were at odds with Pershing's vision of how the AEF was to fight. AEF historian James Rainey has stated that "the tactical doctrine of the AEF was fraught with inconsistencies," and he was nowhere more correct than in the AEF's attempt to deliver that doctrine to its own fighting forces.[70] Pershing and other senior AEF officers were well aware of the dangers of relying on French and British officers to instruct the American divisions and staff officers, but the dearth of trained and experienced American officers left them little choice. Pershing used Allied officers until American officers were available to replace them. He also fought numerous efforts by both French and British commanders to increase interaction between their forces and his. Pershing feared that such involvement "would prevent the final build up of a cohesive, *aggressive*, self-reliant American force," which he believed had to be "depended upon to deliver the final blows necessary to end this war."[71] Pershing later stated that an army trained solely on trench warfare methods "would probably have lacked the *aggressiveness* to break through the enemy's lines and the knowledge of how to carry on thereafter" (emphasis added).[72]

However, due to problems within the American army's training and mobilization system, as well as to external forces, such as the crisis of the German spring offensives, the AEF training system was not able to unambiguously instill its doctrine, or the skills and tools necessary to execute it, in its fighting forces. Colonel Conrad Lanza claimed that because only "conduct-of-fire" was taught at the French firing centers, with artillery tactics not even being addressed, American artillery officers "left for the front without having considered the employment of artillery in battle."[73] That may have been true as far as open warfare methods were concerned, since only a few days were devoted to that form of employment. General Hinds admitted after the war that the training program envisioned by AEF officials, which called for an entire month of open warfare training, was not provided for any artillery brigade except the 1st FA Brigade of the U.S. 1st Division (the only division that completed the entire schedule of training).[74]

Nevertheless, brigades like the 2nd Field Artillery Brigade did not come out of the training empty-handed. While its training regimen may have failed to teach the division what it needed to know to be proficient in open warfare, a substantial body of evidence suggests that it did acquire many of the skills, along with the more conservative artillery-focused doctrine, that would be used to successfully engage in more limited, set-piece attacks. Just what it had learned, and what it had yet to learn, was to be demonstrated on the field of battle.

NOTES

1. John J. Pershing, *My Experiences in the World War* (New York: Frederick A. Stokes Company, 1931), volume I, p. 152.

2. David F. Trask, *The AEF and Coalition Warmaking, 1917–1918* (Lawrence: University Press of Kansas, 1993), p. 4.

3. American Battle Monuments Commission (hereafter ABMC), *American Armies and Battlefields in Europe* (Washington, D.C.: U.S. GPO, 1995, 1938), p. 15.

4. Quoted in Trask, *Coalition Warmaking,* p. 5.

5. Rod Paschall, *The Defeat of Imperial Germany, 1917–1918* (New York: Da Capo Press, 1994), pp. 37–38; Braim, *The Test of Battle,* p. 31–32; Zabecki, *Steel Wind,* p. 111; Colonel Henry J. Reilly, "Fontainebleau in War Time," *Field Artillery Journal* (hereafter *FAJ*) (April–June, 1917), p. 113.

6. Braim, *Test of Battle,* pp. 32; Paschall, *Defeat of Imperial Germany,* pp. 48–49. For more information on the French mutinies following the Nivelle Offensive, see Richard M. Watts, *Dare Call It Treason* (New York: Simon and Schuster, 1963).

7. U.S. War Department, *Order of Battle of the United States Land Forces in the World War,* Volume 3, Zone of the Interior (Washington, D.C.: U.S. GPO, 1988), pp. 199–201.

8. Donald Smythe, *Guerrilla Warrior* (New York: Charles Scribner's Sons, 1973), p. 222; Russell F. Weigley, *History of the United States Army* (Bloomington: Indiana University Press, 1967), p. 351.

9. Editorial Department, *FAJ* (January–March, 1917), 71–72.

10. Conrad H. Lanza, "The Artillery Support of the Infantry in the AEF," *FAJ* (January–March, 1936), p. 62.

11. War Department, *Provisional Drill and Service Regulations for Field Artillery (Horse and Light), 1916, corrected to April 15, 1917* (New York: Military Publishing Co., 1917), pp. 9, 178.

12. During the nineteen-day artillery preparation for the British Passchendaele Offensive near Ypres in July 1917, Allied gunners shot 4,300,000 rounds at the defending Germans. Paul Kennedy, "Britain in the First World War," in *Military Effectiveness,* Volume I: *The First World War,* edited by Allan R. Millett and Williamson Murray (Boston: Allen and Unwin, 1988), p. 55.

13. "Notes on Artillery," *FAJ* (April–June, 1917), pp. 179–80. For more detailed discussions of this doctrine, see J. B. A. Bailey, *Field Artillery and Firepower* (Oxford: The Military Press, 1989) and Zabecki, *Steel Wind,* Chapter 2.

14. The development of prewar artillery doctrine in the U.S. Army is dealt with in an unpublished MA thesis by Steven A. Stebbins, "Indirect Fire: The Challenge and Response in the U.S. Army, 1907–1917" (Chapel Hill: University of North Carolina, 1993), pp. 79–93.

15. "New Field Artillery Classification," *FAJ* (January–March, 1917), p. 29. Some of the "new methods" that Aultman referred to were the increased use of indirect fire and the use of new techniques to improve accuracy, such as applying

corrections for atmospheric conditions, changes in muzzle velocity due to wear, and variances in the powder used in different lots of shells. These were all detailed in "Current Field Artillery Notes," *FAJ* (April–June, 1917), pp. 198–204.

16. Pershing, *My Experiences*, I: 38.

17. U.S. Department of the Army. *United States Army in the World War, 1917–1919* (hereafter *USAWW*), 17 vols. (Washington, D.C.: U.S. GPO, 1948), vol. 1, p. 4; *USAWW*, 1: 93.

18. For a complete display of the AEF's organization, see the numerous "Tables of Organization" in *USAWW*, 1: 155–423.

19. James G. Harbord, *The American Army in France, 1917–1919* (Boston: Little, Brown, 1936), p. 103.

20. James W. Rainey, "Ambivalent Warfare: The Tactical Doctrine of the AEF in World War I," *Parameters*, Journal of the U.S. Army War College (September, 1983), p. 40.

21. The report on the meeting between the Baker Mission and the AEF staff is in *USAWW*, 1: 109–14.

22. Fully-manned British divisions had approximately 15,000 men and were at various times supported by two to three organic twenty-four gun "brigades" of artillery, or between forty-eight and seventy-two guns. Eventually, the British settled on giving each division forty-eight organic guns. The 12,000-man French division, which probably had about one third of the infantry of a full AEF division, was supported by one thirty-six gun "regiment." *USAWW*, 1: 68; Harbord, *American Army*, p. 101; Donald Smythe, *Pershing: General of the Armies* (Bloomington: Indiana University Press, 1986), p. 36.

23. *USAWW*, 1: 68.

24. Summerall demonstrated how he thought American artillery forces should fight by describing how they should be trained. His training suggestions read like a trench warfare artillery skills listing: emphasis on "accuracy of fire, meteorological corrections, battle charts, map exercises in constructing barrages, adjustment and conduct of fire by aeroplane," and in sound ranging and flash spotting. *USAWW*, 1: 73.

25. The U.S. effort to provide the AEF with American-built guns was a complete failure; the AEF was forced to fight the entire war with the guns it could get from the French and British. No American field artillery regiment that saw combat ever received an American-built gun before the war ended. See Ivan V. Hogg, *The Guns, 1914–1918* (New York: Ballantine Books, 1971).

26. For more information on these weapons see Hogg, *The Guns*, and Gudmundsson, *On Artillery*.

27. "The French Government was to furnish the guns necessary for 'all divisional artillery arriving in France,' " *Report, Chief of Artillery, AEF*, Inclosures-Part IV, "Material," p. 1, Folder 382, Commander-in-Chief Reports, Entry 22, RG 120, NA. It is hard to overemphasize the importance of this decision; as Frank Comparato wrote, "With the almost complete adoption of French ordinance by American artillery, the position of France in artillery matters was supreme." See Comparato, *Age of the Great Guns*, p. 52.

28. *Report, Chief of Artillery, AEF*. Inclosures-Part II, "Field Artillery Training," p. 1, Folder 382, Commander-in-Chief Reports, Entry 22, RG 120, NA.

29. The Army War College prepared seventy-one foreign documents for pub-

lication between 23 May and 31 August 1917. Fifteen of these dealt specifically with artillery and numerous others addressed the use of artillery in more general terms. Stebbins, "Indirect Fire," p. 85.

30. Army Service School, *Instructions for the Training of Divisions for Offensive Action* (Ft. Leavenworth, KS: Army Service Schools Press, 1917), pp. 12–15.

31. William J. Snow, *Signposts of Experience* (Washington, D.C.: U.S. Field Artillery Association, 1941), pp. 41–42.

32. "The Mission of the School of Fire for Field Artillery: Address to Incoming Class by Commandant, Col. A. S. Fleming," *FAJ* (October–December, 1917), pp. 384–87.

33. *Report of G-5*, Appendix 31, "Divisional Training," pp. 6–7, Folder 246, Commander-in-Chief Reports, Entry 22, RG 120, NA; *Report, Chief of Artillery, AEF*, Inclosures-Part II, "Field Artillery Training," p. 8, Folder 382, Commander-in-Chief Reports, Entry 22, RG 120, NA.

34. Commenting after the war on the Passchendaele Offensive, Pershing wrote, "The theory of winning the war by attrition, with isolated attacks on limited fronts, which was evidently the idea of the British General Staff, did not appeal to me in principle." Pershing, *My Experiences*, I: 114; Harvey A. DeWeerd, *President Wilson Fights His War* (New York: Macmillan, 1968), p. 191.

35. Pershing, *My Experiences*, II: 237.

36. Pershing, *My Experiences*, II: 12.

37. War Department, *Field Service Regulations: United States Army, 1914, Corrected to July 31, 1918* (Washington, D.C.: GPO, 1918), pp. 95–96.

38. Quoted in Rainey, "Ambivalent Warfare," p. 45.

39. Harbord, *American Army*, p. 149.

40. See John J. Pershing, *Combat Instructions*, AEF Document No. 1348 (France: G-5 Section, GHQ, AEF, 5 September 1918), p. 1; Memo, No. 52, HQ 2nd Division, 13 May 1918, 2nd Division Historical File, Entry 1241-50.4, RG 120, NA.

41. For Nivelle's plans for his 1917 offensive see Watt, *Dare Call It Treason*, Chapter 10. See Tim Travers, *The Killing Ground* (London: Unwin Hyman, 1987), pp. 203–17 for an analysis of Haig's plans at Passchendaele.

42. Pershing, *My Experiences*, II: 237; *Report of G-5*, Appendix 31, "Divisional Training," pp. 7–8, Folder 246, Commander-in-Chief Reports, Entry 22, RG 120, NA.

43. *Field Service Regulations*, p. 74; *Report of G-5*, Appendix 31, "Divisional Training," p. 7, Folder 246, Commander-in-Chief Reports, Entry 22, RG 120, NA.

44. Cablegram No. 228-S, 19 October 1917 in *Report, Chief of Artillery, AEF*, Inclosures-Part II, "Field Artillery Training," p. 15, Folder 382, Commander-in-Chief Reports, Entry 22, RG 120, NA.

45. Cablegram No. 952-S, 18 April 1918 in *Report, Chief of Artillery, AEF*, Inclosures-Part II, "Field Artillery Training," p. 15, Folder 382, Commander-in-Chief Reports, Entry 22, RG 120, NA.

46. Lanza, "Artillery Support," p. 73.

47. Pershing, *My Experiences*, I: 150.

48. Department of Gunnery, School of Fire for Field Artillery, "American Drill Regulations and 'Artillery Firing' " *FAJ* (July–September, 1918), p. 364.

49. Pershing, *My Experiences*, I: 152.

50. The history of the 12th Field Artillery described the time spent in the

United States before sailing to France as "typical garrison life in the United States," an indication of the poorly coordinated training that existed in the United States until well into 1918. *History of the 12th Field Artillery*, no author, unpaginated typescript in Folder 16, 2nd Division files, G-3 Reports, AEF/GHQ, Entry 270, RG 120, NA; Oliver L. Spaulding and John W. Wright, *The Second Division, American Expeditionary Force in France, 1917–1919* (New York: The Hillman Press, 1937), p. 6–7.

51. Ironically, this group of officers received perhaps the most intensive artillery training in France. Captain Clark reported that he arrived at Valdahon after the 1st FA Brigade had finished there and before the 2nd Brigade arrived. He joined about forty other American officers who received virtual "one on one training sessions" with the French instructors for a whole month until the 2nd Brigade arrived from the states. He and the other junior officers were then able to help train the American artillerymen on the new guns. Captain John D. Clark, "My War," unpublished memoir in 15th FA folder, 2nd Division boxes, World War I Manuscript Collection at the U.S. Army Military History Institute (USAMHI).

52. An analysis of 2nd FA Brigade survey responses in the World War I Manuscript Collection at USAMHI confirms that neither officers nor gunners had even seen their French weapons until reaching the Valdahon camp; *History of the 12th Field Artillery*, no author, unpaginated typescript in Folder 16, 2nd Division files, G-3 Reports, AEF/GHQ, Entry 270, RG 120, NA.

53. George L. Irwin, "Notes on the Training and Handling of Divisional Artillery in France," *FAJ* (November–December, 1919), p. 489.

54. The 15th FA (75mm), commanded by Colonel Thomas E. Merrill, and 17th FA (155mm), commanded by Colonel Albert J. Bowley, both arrived in France on 27 December 1917 and began training on 14 January 1918. The 12th FA (75mm), Colonel Manus McClosky in command, arrived on 25 January and began training on 1 February. All three regiments finished the first phase by 18 March 1918. The 2nd FA Brigade was first commanded by Brigadier General William J. Kenly, who was replaced by Brigadier General George L. Irwin on 1 February 1918. Lieutenant William D. Bickham and Major William E. Burr, "Second Field Artillery Brigade History," p. 1, undated typescript in 2nd FA Brigade Historical File, Entry 1246-11.4, RG 120, NA: *Report, Chief of Artillery, AEF*, Inclosures-Part II, "Field Artillery Training," p. 122, Folder 382, Commander-in-Chief Reports, Entry 22, RG 120, NA.

55. *Report, Chief of Artillery, AEF*, Inclosures-Part II, "Field Artillery Training," pp. 25, 38, Folder 382, Commander-in-Chief Reports, Entry 22, RG 120, NA.

56. *Report, Chief of Artillery, AEF*, Inclosures-Part II, "Field Artillery Training," pp. 1–2, 39, Folder 382, Commander-in-Chief Reports, Entry 22, RG 120, NA.

57. Lieutenant William D. Bickham and Major William E. Burr, "Second Field Artillery Brigade History," pp. 2–3, undated typescript in 2nd FA Brigade Historical File, Entry 1246-11.4, RG 120, NA.

58. Irwin, "Notes on Training," p. 489.

59. Irwin, "Notes on Training," p. 489.

60. The first of the German offensives, near the town of St. Quentin, was launched on 21 March 1918, and the second German drive, the Lys Offensive near Ypres, began on 9 April 1918.

61. A memorandum for Pershing's chief of staff, dated 30 January, stated that after the 2nd Division finished its tour at the front "not later than April 10," it was to be allowed "not less than one month to complete its divisional training," 2nd Division Historical File, Entry 1241-50.4-56.3; Lieutenant William D. Bickham and Major William E. Burr, "Second Field Artillery Brigade History," pp. 4–7, undated typescript in 2nd FA Brigade Historical File, Entry 1246-11.4, RG 120, NA; *History of the 12th Field Artillery*, no author, unpaginated typescript in Folder 16, 2nd Division files, G-3 Reports, AEF/GHQ, Entry 270, RG 120, NA.

62. The division suffered a total of 838 casualties during its two-month tour in the Verdun, Toulon, and Tryon sectors, Report of 2nd division commander (Major General John A. Lejeune) to assistant chief of staff, G-3, GHQ, AEF, Subject: "Operations Information," 30 December 1918, in *Reports and Other Records Regarding the History of AEF Units*, Divisional Skeleton Histories, Entry 443, RG 120, NA.

63. Pershing, *My Experiences*, I: 381–82, II: 48.

64. Letter from General Pershing to chief of French Mission, 16 May 1918, 2nd Division Historical File, Entry 1241-50.4-56.3, RG 120, NA.

65. Lieutenant William D. Bickham and Major William E. Burr, "Second Field Artillery Brigade History," pp. 4–7, undated typescript in 2nd FA Brigade Historical File, Entry 1246-11.4, RG 120, NA; John A. Lejeune, *The Reminiscences of a Marine* (Philadelphia: Dorrance and Co. 1930), p. 291.

66. Lecture 28, "Artillery III: The Use of Artillery in the Offensive," 24 December 1917, 1st Course, Army General Staff College (AGSC), AEF, G-5 Schools File, Entry 362, RG 120, NA.

67. Lecture 20, "Division in the Offensive (I)," 19 December 1917, 1st Course, AGSC, G-5 Schools File, Entry 362, RG 120, NA; Field Service Regulations, p. 91.

68. Conference 107, "Exploitation of Success," 29 April 1918, 2d Course, AGSC, G-5 Schools File, Entry 362, RG 120, NA.

69. Lecture 22, Colonel Koechlin-Schwartz (French), "Artillery in War of Movement," 2 May 1918, 2nd Course, AGSC, G-5 Schools File, Entry 362, RG 120, NA.

70. Rainey, "Ambivalent Warfare," pp. 34–35.

71. The French were well aware of the doctrine that Pershing wanted to use as his guide, and they did not think much of it. A 1 May 1918 memorandum from the French GHQ to all French officers on duty with American troops stated that "Americans dream of operating in open country after having broken through the front. This results in too much attention being devoted to this form of operations." Pershing, *My Experiences*, II: 36; Letter, General Pershing to General Pétain, in *Report of G-5*, Appendix 31, "Divisional Training," Folder 246, Commander-in-Chief Reports, Entry 22, RG 120, NA.

72. Pershing, *My Experiences*, I: 152.

73. Lanza, "Counterbattery in the AEF," *FAJ* (July–September, 1936), p. 457.

74. Major General Ernest Hinds, Lecture to the Army Center of Artillery Studies (hereafter ACAS), "The Training of Artillery in France," 18 December 1918, p. 4, ACAS Files, Entry 371, RG 120, NA.

2

Belleau Wood: Hard Lessons in Position Warfare

> Again, was decisively shown the great importance of artillery to Infantry [*sic*]. Infantry alone without material, makes little or no progress . . . This lesson was well impressed on the 2nd Division at this time.[1]
>
> Major General John A. Lejeune
> Commanding General, 2nd Division

On 31 May, as units of Major General Omar Bundy's 2nd Division were racing towards the advancing German forces near Chateau-Thierry, French citizens in Paris panicked. Since the beginning of its offensive in the Chemin des Dames area on 27 May, the German Army had advanced more than sixty kilometers and was moving about as fast as its infantry could march. With the German Army less than seventy kilometers from Paris, thousands of civilians evacuated the city, and the French government prepared to move to Bordeaux. General Henri-Philippe Pétain had sent sixteen French divisions from his general reserve towards the advancing Germans, but they had little effect. A French staff officer lamented that the French divisions sent in to stop the German forces "evaporated immediately like drops of rain on white-hot iron" in the face of the German onslaught. Such was the situation surrounding the combat debut of the American 2nd Division (see Map 2.1).[2]

The fighting of the 2nd Division near Chateau-Thierry, in what was known as the Aisne Defensive, and later near Soissons in the Aisne-Marne counteroffensive, proved to be among the most important operations of the war. Near Chateau-Thierry the division demonstrated to

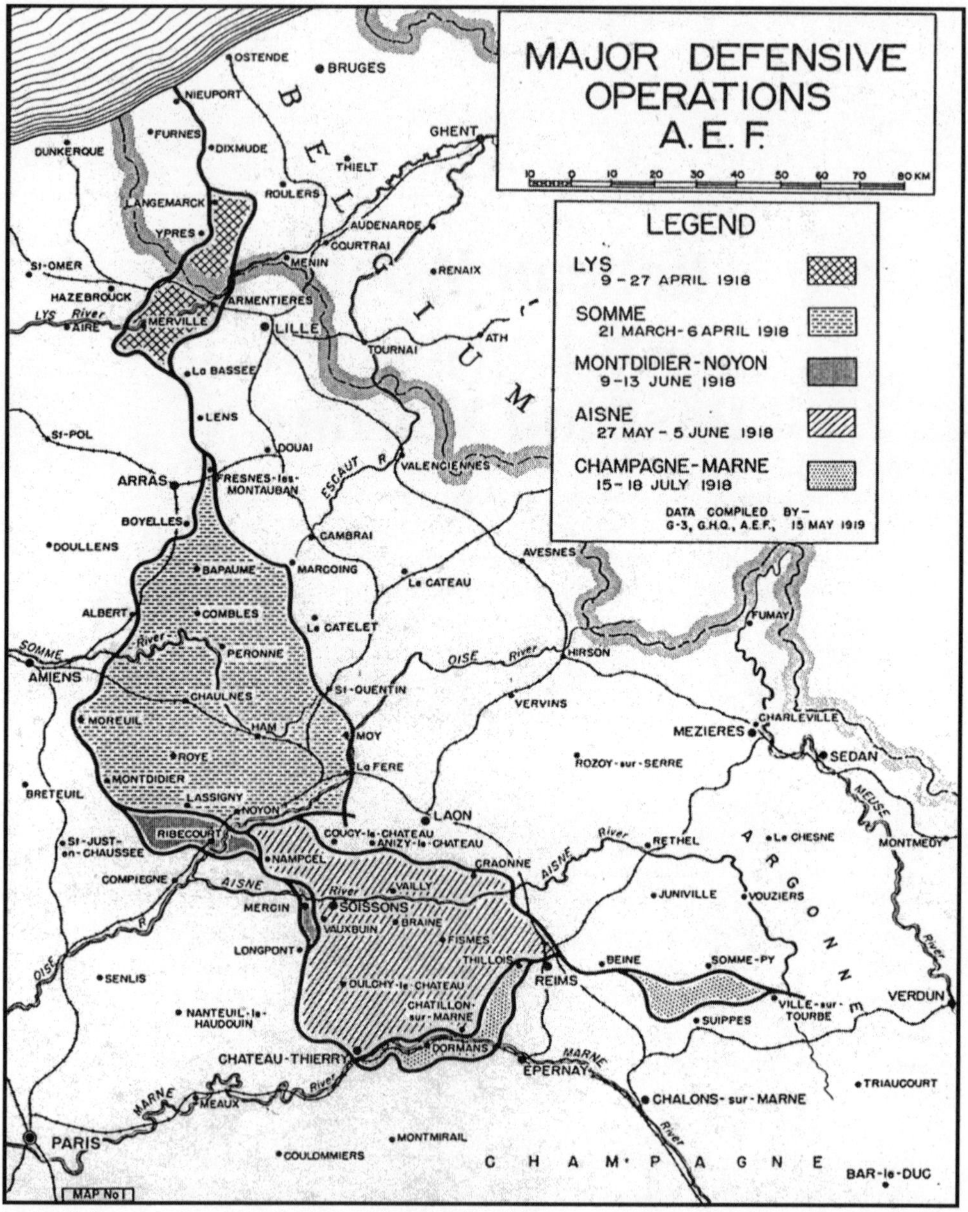

Map 2.1. Significant tactical and operational success of Germany's spring offensives in 1918. Notice that the greatest depth of the German Aisne or Chemin des Dames offensive was just west of Chateau-Thierry, where the 2nd Division fought to take Belleau Wood. Taken from *United States Army in the World War,* vol. 1, p. 15.

both the Germans and the Allies that the Americans were capable of fighting tenaciously. The Allies, and especially the French, were elated and inspired by the timely, impressive showing of the new American division. And the German Army had to adjust its opinion of the impact that American forces might have on the Western Front. Later in July, the division's successful attack near Soissons was an important part of what many military analysts and historians consider to be the turning point of the war.

However, the 2nd Division paid an exceptionally high price for those moral and strategic victories. During the two engagements, the division suffered more casualties than its entire authorized infantry strength, and the manner in which the division utilized its artillery forces was a primary cause of the excessive losses. There were, of course, many external factors that affected the combat effectiveness of the 2nd Division, and especially its use of artillery, during these first operations. While these factors will be discussed in this chapter and the next, it will also be quite clear that the division did not demonstrate the will or the skill to maximize the impact of its supporting artillery forces in these first operations. In short, the division learned some hard lessons about the importance of overwhelming firepower to any infantry attack.

After a trying movement to the combat area by truck and marching, the infantry regiments of the 2nd Division eventually established a line of resistance on 1 June that blocked the Chateau-Thierry-to-Paris highway, approximately ten kilometers west of Chateau-Thierry. While some fragmented French forces held loosely organized positions between the 2nd Division and the advancing Germans, no established French position existed in front of the Americans. The General Operations Order of 1 June for the French XXI Corps, to whom the 2nd Division was then assigned, stated that the so-called American "line of support" was "nearly in contact with the enemy."[3] On the following day localized German attacks sent a number of retreating French troops through the American lines. Some French officers advised and even attempted to order American officers to immediately withdraw; none did. By 3 June, the 3rd and 4th Brigades held a line that stretched for some twenty kilometers, and the divisional artillery batteries had moved into positions behind them.[4] Brigadier General William Chamberlaine, the commander of the 2nd FA Brigade, interviewed the chief of artillery of the French division that had official control of the sector, and made dispositions to defend the American positions, which were rapidly becoming the new front line of the sector. Chamberlaine's staff arranged for the French Army to supply all the ammunition required by the brigade. Chamberlaine then assigned his light regiments to cover specific sectors of the front (generally, the 12th FA Regiment covered the 4th Brigade on the left and the 15th FA Regiment supported the 3rd Brigade on the right),

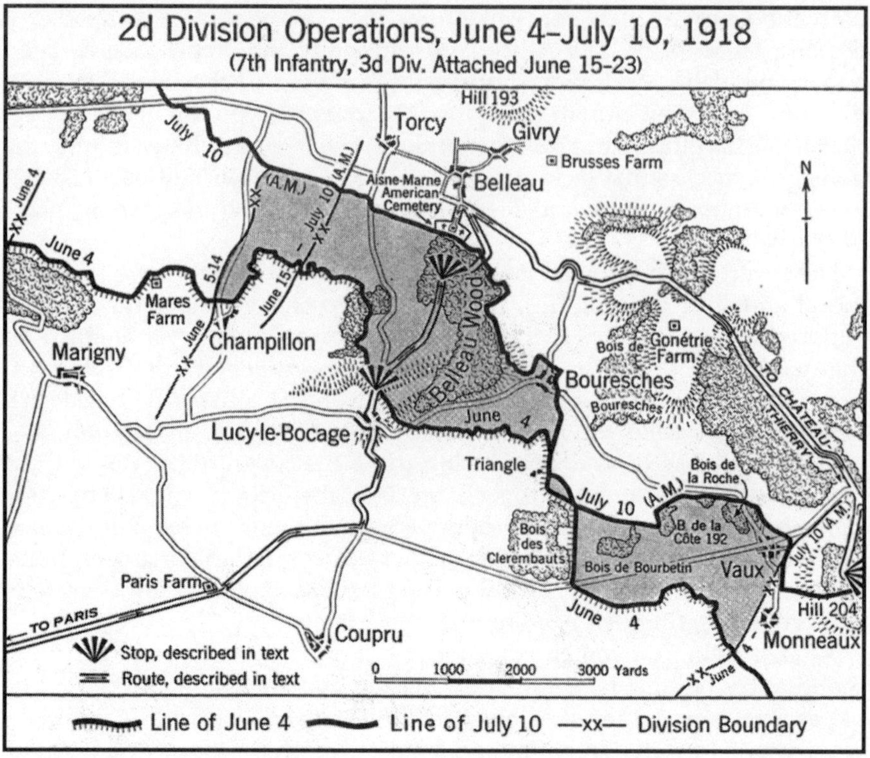

Map 2.2. 2nd Division operations at Belleau Wood and Vaux, 1 June–10 July 1918. Taken from ABMC, *American Armies and Battlefields in Europe,* p. 45.

ordered his regimental commanders to immediately confer with the appropriate infantry commanders to establish defensive support agreements and sent an artillery liaison officer to each infantry battalion in the line.[5]

On the morning of 4 June all remaining French troops in front of the American position were relieved and they passed through the 2nd Division's lines (see Map 2.2). Their supporting French artillery left the sector with them, with the exception of one regiment of 75mm guns. The sector belonged exclusively to the 2nd Division, and it was expected to hold it at all costs.[6] On that day the Marine brigade, on the division's left, received and successfully repulsed a "violent attack." Though the Americans did not know it then, it was to be the final attack of the German offensive. On that day the German IV Reserve Corps, opposite the 2nd Division, announced to its front-line divisions that they were to consolidate their gains, only make additional advances that improved their current line, and "temporarily" assume a defensive posture.

Unaware of this change, General Harbord, the commander of the 4th

Figure 2.1. Major General James G. Harbord. He commanded the 4th Brigade at Belleau Wood and the entire 2nd Division during the Aisne-Marne counteroffensive near Soissons. Many of his attacks, based on Pershing's concept of "self-reliant" infantry, led to heavy casualties. Courtesy U.S. Army Military History Institute.

Brigade (see Figure 2.1), expected the attacks to continue and ordered patrols sent out to the front to secure information on the enemy.[7] The commander of the French XXI Corps, General Joseph Degoutte, likewise feared that the German attacks would continue. Concerned that the present line held by his French and American troops allowed the Germans to discreetly mass troops for future attacks, he ordered the French 167th Division and the U.S. 2nd Division to make short advances to better positions on 6 June.[8] With the exception of the single-day assault and capture of the town of Cantigny by one regiment of the U.S. 1st Division on 28 May, this would be the first offensive action of any kind by an American army unit in the war.

The significance of the upcoming clash between the 2nd Division and

German forces was not lost on either the German or the American commanders. When General Erich von Ludendorff, the de facto commander of the German Army, learned that American soldiers opposed his forces near Chateau-Thierry, he ordered that they "should be hit particularly hard," and prevented from achieving any success whatsoever. General Böhm, the commander of the German 28th Division, located directly across the lines from the 2nd Division, warned his troops:

> Should the Americans on our front even temporarily gain the upper hand, it would have a most unfavorable effect for us as regards the morale of the Allies and the duration of the war. In the fighting that now confronts us, we are not concerned with the occupation or the non-occupation of this or that unimportant wood or village, but with the question as to whether the Anglo-American propaganda that the American Army is equal to or even superior to the German, will be successful.[9]

If anything, the Americans seemed even more concerned about the meaning of their first clash with the enemy. General Harbord insisted that "every man in the Marine Brigade realized that America was on trial as to the courage and fighting quality of her sons." He also echoed the thoughts of the opposing German commander when he stated that "more than the Bois de Belleau [a small wooded area under German control] was at stake . . . it was a struggle for psychological mastery."[10] General Bundy stated that "effects on both American and German morale and prestige . . . make imperative the occupation of the Bois de Belleau."[11] In this emotionally and psychologically charged environment, the senior officers in the 2nd Division believed the unit had to demonstrate every bit of aggressiveness that it could muster.

Orders called for the first attack to take place at 3:45 AM, on 6 June. General Harbord ordered one battalion of the 5th Marine Regiment, Major Julius S. Turrill's 1st Battalion, as well as two machine-gun companies, to advance the division's left flank about one kilometer and take possession of a small elevation called Hill 142. This would keep the division's left flank in touch with the French 167th Division on the left, which was to advance its entire line to a better position. Major Benjamin S. Berry's 3rd Battalion of the 5th Marine Regiment, on Turrill's immediate right, was to advance his left to stay in contact with Turrill's unit.[12]

In accordance with the approved American doctrine, artillery support for the advance was slight. Between the French and American artillery regiments in the sector, six batteries of 75s and two batteries of 155s were available for the operation. However, the plan called for only minor use of these guns. The orders clearly stated that "there will not be any preparation properly speaking, so as not to attrack [*sic*] the attention of the enemy." The division would carry out some "raking fire" on Hill 142 during the preceding night, but only with the 75s. During the attack the howitzers would focus exclusively on the interdiction of rear areas. For five minutes before the infantry assault, the 75s were to place "a violent annihilating fire" on the first objective, but then lift their fire to more distant targets just before the Marines left their attack positions. General Chamberlaine charged Colonel Manus McClosky, the commander of the 12th FA, with coordinating all other details of the artillery support plan with General Harbord.[13] Apparently, Harbord made no further requests.

The attack began right on time. By 7:00 AM, Majors Turrill and Berry had taken their objectives and were consolidating their new positions. However, they reported that they had advanced through "heavy shell and rifle fire" and had suffered a number of casualties. Nevertheless, even before exact casualty figures were reported, the attack was treated as a great success, and congratulatory messages were sent out to the appropriate regimental and battalion commanders.[14]

General Degoutte's corps orders had instructed that "as soon as possible" after the first attack, the 2nd Division was to make a second advance to take control of Belleau Wood (the Bois de Belleau), a small kidney-shaped wood in front of the 4th Brigade. In compliance with this directive, General Bundy directed the 4th Brigade to capture the wood in an attack scheduled for 5:00 PM that afternoon. General Harbord issued the necessary attack order to his troops at 2:00 PM. Plans directed four battalions from the two Marine regiments to attack in two phases, with the first phase seizing the wood and the second taking the town of Bouresches, just east of the southern edge of the wood.[15]

While the division had nearly twelve hours to prepare for the morning attack on 6 June, it had just three hours to make all arrangements for the afternoon assault, which was larger and much more complex. Strangely, the artillery support plan for the afternoon attack called for even less actual fire support than had been provided for the morning attack. Though more guns were used, the fire plan called only for "interdiction and harassing fires in the zone to be attacked," as well as for some fire "on the points where it is possible that there are nests of machine guns." There was no specification as to the duration or rate of fire for the preparation, nor was there a call for any type of strong barrage to aid the infantry in its advance.[16] While this weak fire-support plan might have been the result of the brief time allowed to coordinate the artillery effort, it nevertheless appears that General Harbord intended for his Marines to make the attack with a minimal amount of artillery support.

Harbord later claimed that French commanders had informed him that "Belleau Wood was not occupied, except by a very short line across the northeast corner which was entrenched." Apparently relying on this information, which could not have been any more recent than the early morning of 4 June, General Harbord called for his Marines to make "a surprise attack," which would not be supported by "any unusual artillery activity." This attack turned out to be a test case of Pershing's theories about the "self-reliant" infantryman. The result was a tragedy, for as Harbord was to later admit, "the Bois proved to be very occupied, with many machine gun nests, in positions well chosen among giant boulders."[17]

At 5:00 PM the infantry started forward. An observer reported that at least one battalion advanced "in beautiful deployment, in beautiful line." This might have described the attack of Major Berry's battalion, which had to advance across a wide open field before reaching the west edge of Belleau Wood. When the battalion was half way to the wood, German machine guns opened up and shredded the attacking waves to pieces. When Colonel Wendell Neville, the commander of the 5th Marine Regiment, reported on Berry's advance, he claimed that they probably never even reached the woods and that most of the battalion was either killed or wounded.[18]

The situation was only a little better for Major Berton W. Sibley's 3rd Battalion of the 6th Marine Regiment, which hit a machine-gun nest shortly after entering the southern part of the wood. Sibley reported that the machine guns "had done a lot of damage" to his unit, and he said he was "unable to advance" further due to the strength of the defensive position. Lieutenant Colonel Harry Lee, the acting commander of the 6th Regiment, told General Harbord that it "seems impossible to attack hostile [machine] gun positions without artillery." And that is exactly what Harbord directed the Marines to do that day. The fighting in and around Belleau Wood raged into the night, until Harbord finally ordered his men to stop attacking and to consolidate their positions.[19] The Marines suffered 1,087 casualties in the two attacks of 6 June; it proved to be the bloodiest day in Marine Corps history, and would remain so throughout the war and well into World War II.[20] The only successes of the attack were the toehold gained in the southern edge of the wood and the successful seizure of the town of Bouresches on the right flank.

Unfortunately, Major Sibley's battalion attempted a few more localized attacks in the southern edge of the wood on 7 and 8 June; all were unsuccessful. While Harbord did have his artillery "play" on the wood for a brief period prior to one of the attacks, all indications are that the infantry attacked without any coordinated artillery support. By 10:30 AM on 8 June, Sibley reported that his losses were "severe," that the machine-gun nests were "too strong for our infantry," and that stronger artillery support had to be provided prior to any further attacks. After the war, General Lejeune summarized these attacks by directly stating that they were attempted "without artillery preparation." He claimed that in each attack "little progress was made and it became apparent that the reckless courage of the foot soldier with his rifle and bayonet could not overcome

Figure 2.2. Soldiers from the 23rd Infantry Regiment provide fire support with a 37mm gun. This was the most firepower the "self-reliant" infantry had at their disposal. Courtesy U.S. Army Military History Institute.

machine guns, well-protected in rocky nests." [21] These lessons had been learned by numerous British and French soldiers in the previous three years and should have been learned by the Americans before the Marines ever attacked.

By 8 June, nearly everyone in the division, if not the entire French XXI Corps, understood that the Marine attacks of the first two days, based on "self-reliant" infantry, had been too costly (see Figure 2.2). General Degoutte felt compelled to direct that future attacks by the 2nd Division "be conducted methodically, by means of successive minor operations, making the utmost use of artillery and reducing the employment of infantry to the minimum necessary."[22] The attacks from 6 through 8 June were anything but methodical, and they certainly did not maximize the use of artillery.

On 8 June General Chamberlaine proposed to General Harbord a different method of taking Belleau Wood. He suggested that the entire wood be "systematically covered by artillery fire preparatory to another attack." In essence he suggested that Harbord shelve Pershing's infantry-based method of attack in favor of the Allied artillery-centered approach, and Harbord acquiesced. The commanders agreed to make the next attack on 10 June, after a thorough artillery preparation. By mid-day 8 June, Harbord had ordered all the Marines then in Belleau Wood to withdraw to a clearly established line at the southern edge of the wood to permit the artillery "a free hand" in softening up the German machine-gun nests.[23]

With more than a full day to prepare the fire plan, this was to be the first real test of the 2nd Division's ability to plan and execute an artillery-centered attack. Despite the extensive protection that the wood afforded the German strong points, the artillery plan called for just one hour of preparatory bombardment—but it was a very well-detailed artillery preparation that required all the guns in the division sector, over 160 in total, to deliver intense fire on specific points in and around the wood. Both 75s and 155s were ordered to hit certain areas in the wood, while still other 155s were detailed to fire on rear locations. Last, and of great importance, for the first time, the 75s were to fire a real rolling barrage that would walk the infantry step by step through the enemy positions. To inflict maximum casualties on any retreating Germans and to provide an additional measure of protection against a counterattack, the fire plan called for twelve machine guns near Bouresches to fire a barrage just east of the woods. The plan was impressive on paper. What remained to be seen was whether the division could carry it out.[24]

At 3:30 AM on 10 June, the guns of the 2nd FA Brigade let loose on the German positions, depositing some 28,000 rounds of 75mm shells and over 12,000 rounds from the 155mm howitzers into Belleau Wood and the surrounding area (see Map 2.3). An hour later the 75s shifted to a rolling barrage that advanced one hundred meters every two minutes. The barrage was to lead the attacking Marines all the way to their objective in about twenty-two minutes.[25]

Immediately after the Marines started forward, the commander of the attacking battalion, Major John H. Hughes, reported "artillery barrage working beautifully." Unlike the previous attacks, there were only a few short bursts of enemy machine-gun fire during the advance. Right on schedule, Hughes reported taking his objective "without opposition." As for the effectiveness of the preparation and barrage, Hughes exclaimed that the "artillery has blown the Bois de Belleau to mince meat." Colonel Lee stated that by 5:20 AM the attempt to take the southern half of the wood was "deemed finished." Not only was the objective gained, but it was secured at a cost in casualties of just eight killed and twenty-four wounded. The differences between this artillery-centered attack and the

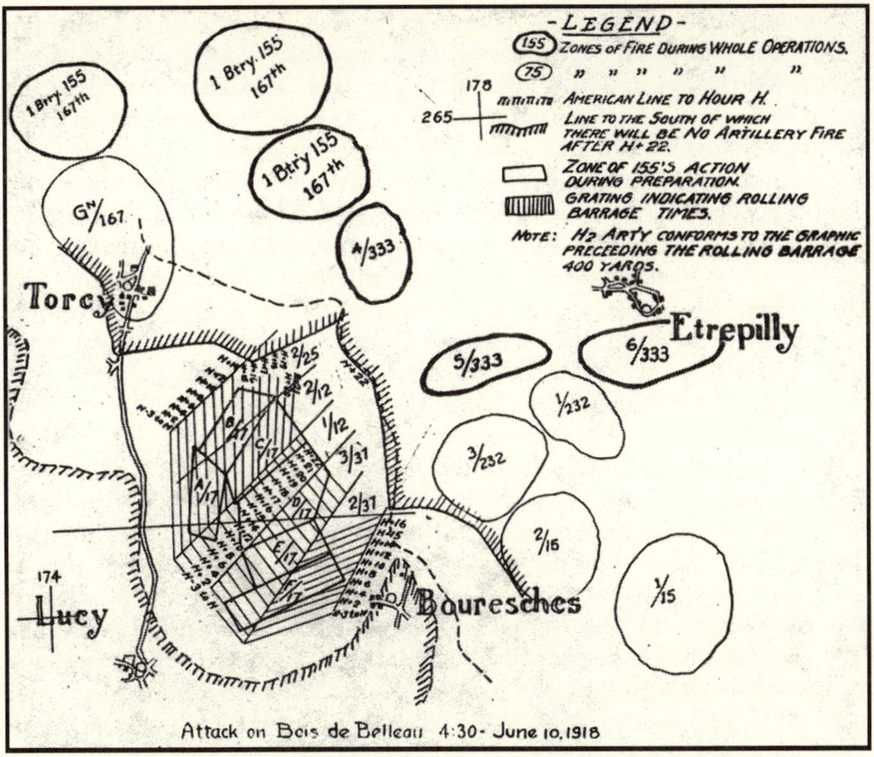

Map 2.3. Barrage chart for the 2nd Division's 10 June 1918 attack on Belleau Wood. Taken from U.S. Army, *Records of the Second Division*, vol. 10.

unsupported infantry assaults of the previous days were obvious, even to Harbord.[26]

That afternoon General Harbord arranged to take the northern half of Belleau Wood with a plan that called for a maximum use of artillery. This time the attack was to come from the west side of the wood, by the 2nd Battalion, 5th Marine Regiment, commanded by Lieutenant Colonel Frederick M. Wise. The attack was an attempt to flank the German positions facing south toward Major Hughes' battalion. Again, the plan directed the artillery to fire a one-hour-long preparatory bombardment of the woods before the infantry attack, which was scheduled to take place at 4:30 AM on the following day. The same number of guns were employed, and in much the same manner.[27]

While this attack was much more successful than those of the first three days, it was not as clear a victory as the 10 June attack had been. The front line was advanced to the northern section of the wood, even reaching the northeast edge at one time, but a number of machine-gun nests had escaped both the artillery preparation and the rolling barrage, and they prevented Colonel Wise from completely clearing the northern half of the wood. His casualties were also much heavier than those of the previous day, in excess of 200. The main reason for the trouble appears to lie in the fact that the attack of 10 June had stopped short of the main defensive line in the wood, and this line, well sheltered by logs and boulders, was not as easy to destroy as the weaker positions in the southern part of the wood. Also, since Major Hughes' battalion maintained its forward position in the wood and was not withdrawn to allow the artillery a "free hand," the artillery placed its preliminary fire and started its rolling barrage too far in advance of the American positions. Colonel Wise reported that when his men advanced, they "found quite a few machine-gun nests inside of the barrage which gave a great deal of trouble." Nevertheless, the Marines advanced the American line and captured over 400 German prisoners, but at a higher cost in casualties than in the previous day's assault.[28]

On 12 June General Harbord held a conference with his regimental and battalion commanders to discuss plans to take the rest of the wood. Colonel Wise stated that he was confident that with some further artillery preparation, his battalion could do the job. Harbord then requested that the 12th FA fire on the wood from 4:00 PM to 5:00 PM, at which time Wise was to make another attack.[29] While no details of the artillery fire plan exist, judging by the results, it clearly was inadequate.

At 4:30 PM, Colonel Wise reported that the artillery preparation was not doing the job. His artillery liaison officer had told him that the preliminary bombardment was "entirely too light," and that whenever there was a lull in the fire, the Germans would come out of their secure positions with their machine guns ready to meet the attack. He requested that the barrage continue for an extra hour, at an increased intensity, and admitted that he feared he was "going to have a little bit of trouble," because the area being shelled was "entirely too large for the number of guns assigned." Harbord granted the request, but probably should have done more, considering the warning signals his battalion commander was sending.[30]

When Wise's battalion attacked, it was able to push through to its objectives and clear out the machine-gun nests in the northern part of the wood, but the casualties were again rather heavy. Between the casualties sustained in the actual advance and those that came afterward when the German artillery pounded the new American positions, Wise's battalion had lost so many men that it was unable to maintain a front line that was strong enough to prevent German troops from sneaking back into the wood. While the divisional report for the operation claimed that the operation was a "complete success," and that the wood was "completely cleaned out of enemy detachments," the inability of the Marines in the wood to maintain proper liaison with each other throughout the forest guaranteed that there would be more fighting in the northern part of Belleau Wood.[31]

Though the advances of 11 and 12 June were not as painless as that of the 10th, their various degrees of success stood in marked contrast to the abject failures of the attacks attempted between 6 and 8 June. Commenting on these attacks after the war, General Lejeune exclaimed, "strikingly obvious is the great need for artillery in the attack, when one contrasts the little progress made without it and the advances of the last two days."[32] However, it is clear that the lesson was not learned by everyone in the AEF, nor even by the most senior officer in the 4th Marine Brigade, General Harbord.

On the night of 17 June, the 4th Marine Regiment was relieved by troops from the 7th Infantry Regiment of the U.S. 3 Division. During their first night in the northern part of the wood, they were greeted by intense machine-gun fire from German troops who had filtered through the thin American line. The Germans had, in fact, constructed a machine-gun nest in the northern part of the wood, and it became obvious that a new attack would have to be launched to clear the wood. On 20 June elements of the 7th Infantry attempted to dislodge the German machine guns without any artillery support whatsoever; the results were predictable. The attack was a complete failure. On 21 June a second attack was made, this time after some artillery preparation as well as a rolling barrage. It was also unsuccessful. The commander of the 7th Infantry reported that the failure was caused by a perfectly timed and placed counterbarrage that separated his infantry from their own rolling barrage. General Harbord suspected that the failure was caused more by the ineffectiveness of the officers and troops involved. The next day he ordered his Marines back into the wood to do the job.[33]

Despite the successes of the artillery-centered attacks on 10, 11, and 12 June, and the costly failure of every other attack, General Harbord inexplicably returned to a more infantry-focused approach to clear the wood. Although divisional intelligence reports estimated that there were as many as 200 German troops and 15 machine guns in the northern part of the wood, Harbord reverted to his initial infantry-focused methods. On 21 June, he sent the 3rd Battalion of the 5th Marine Regiment back in to clear the wood. Harbord told the new battalion commander, Major Maurice E. Shearer, to have the wood cleared by the next day, and that it was "not practicable to withdraw again and give further artillery preparation." Harbord asserted that "by judicious use of sharpshooting snipers you can reduce the German position without much expenditure of men. . . . With the sniping . . . you should be endeavoring to get the machine gun nests surrounded so you can rush them when ready."[34]

At 7:00 PM on 23 June Major Shearer attempted to take the enemy machine-gun positions without artillery support, just as Harbord had ordered. Though the initial reports spoke of machine guns being captured and only "slight casualties," the reality was much worse. After meeting heavy machine-gun fire and suffering severe casualties, the Marines had to return to their initial positions. According to the regimental commander, Colonel Neville, one of his companies was "almost wiped out." By late morning on 24 June, Major Shearer submitted his report on the operation. In it he stated that he was of the "opinion that infantry alone cannot dislodge enemy [machine] guns."[35]

After this report, General Harbord returned to the only method that had achieved success. He held a conference with all his infantry and artillery commanders, and a detailed attack plan was decided on. In fact, the plan called for more artillery support than had been used in any previous attack. Harbord ordered all Marines in the northern part of the wood to withdraw to an identifiable position near the wood's center by 3:00 AM on 25 June. Beginning at 3:00 AM and continuing for the next fourteen hours, the divisional guns, both 75s and 155s, were to execute a systematic bombardment of the northern section of the wood (see Map 2.4). During the last hour, all guns would fire at their maximum rate, then shift to a rolling barrage that would advance through the wood at a rate of 100 meters every three minutes.[36]

The artillery preparation and attack went off as planned. By daylight on 26 June, Major Shearer, who led the attack, reported to General Harbord, "Woods now US Marine Corps entirely." This time, it stayed that way. While American casualties were initially reported as "somewhat heavy," a final count identified just 123 U.S. total casualties. At that cost the assault netted over 300 German troops and 19 machine guns. Again, the attack with extensive artillery support succeeded where the "self-reliant" infantryman had failed, and all too often, died.[37]

General Lejeune insisted that the 2nd Division learned a costly but valuable lesson in its attempts to take Belleau Wood, namely, "the great importance of artillery to Infantry [sic]."[38] Before summing up the operations around Chateau-Thierry, one more attack deserves mention, the attack of Vaux village, as it provides the best proof that the division did indeed learn how critical overwhelming artillery support was for any infantry attack. The Vaux attack also serves to demonstrate the level of expertise that the division had achieved in its first series of operations against an active enemy.

Five days after the 4th Brigade succeeded in gaining complete control of Belleau Wood, the 3rd Brigade, on the right of the divisional front, attacked and took possession of the town of Vaux, which had been under German control since the 2nd Division took over the front line. What stands out, especially in contrast to the manner in which the division attempted to take Belleau Wood, is how methodically and effectively the division planned and executed the Vaux operation. While the Marines

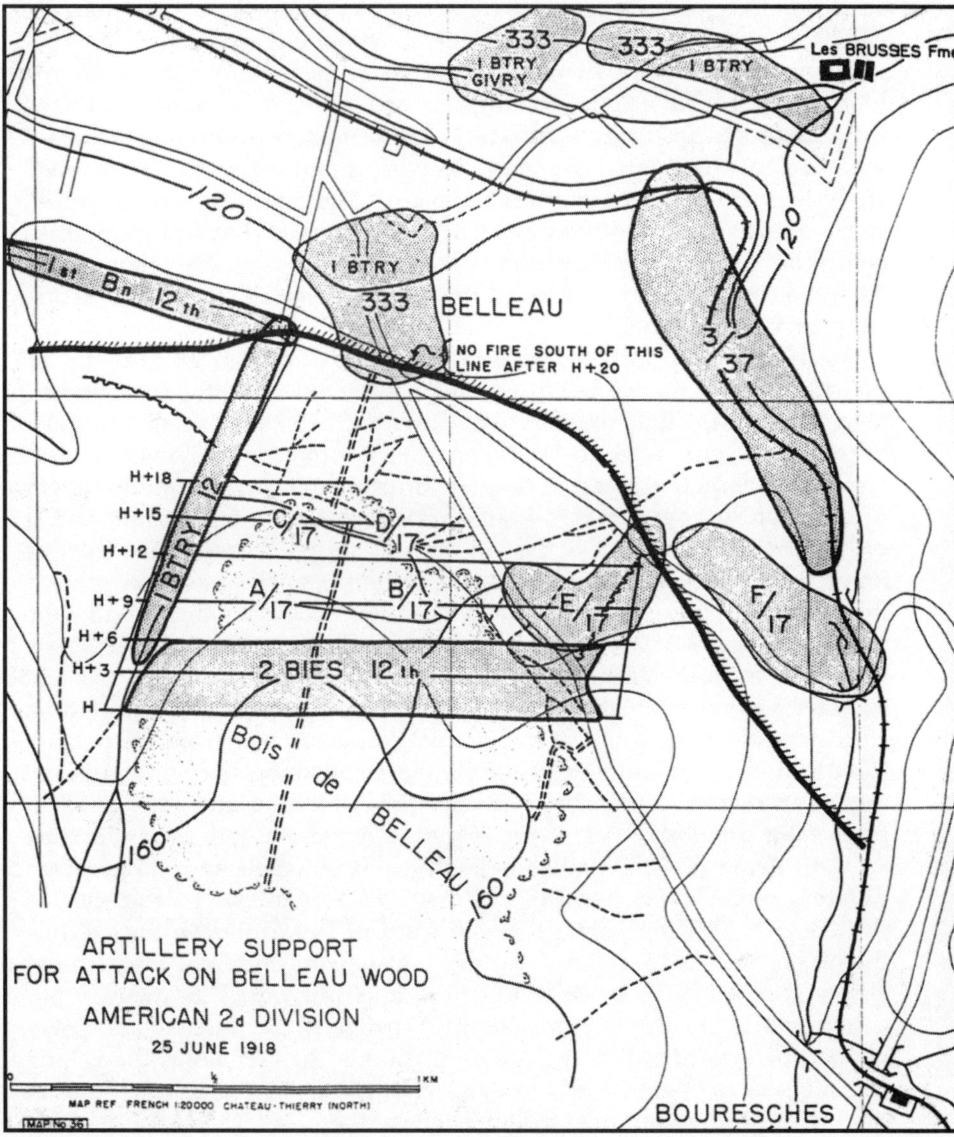

Map 2.4. Barrage chart for the 2nd Division's 25 June 1918 attack on Belleau Wood. Taken from *United States Army in the World War*, vol. 4, p. 561.

had rushed into Belleau Wood not knowing the strength of the German positions there, and with the assumption that the rifle-brandishing infantrymen would be able to handle whatever situation arose, the division successfully seized Vaux after a detailed examination of the German positions and after the coordination of overwhelming artillery support.

Preparations for the attack on Vaux began days before the operation took place. The divisional intelligence and operations sections utilized aerial photographs of Vaux, as well as information from French citizens who had just evacuated the town, to perform an exceptionally detailed analysis of the German defenses in the area. Intelligence troops produced various maps that identified every structure in the town and pinpointed nearly all defensive strong points (see Maps 2.5 and 2.6). From these data, the division staff developed the detailed plan of attack and the fire support arrangements.[39]

For artillery support, the plan incorporated fire from all guns in the divisional sector that were within range of the town and the surrounding area. This meant that sixty-six 75s, thirty-six 155s, as well as a number of trench mortars, were to fire on an area less than two kilometers wide and two kilometers deep. The plan directed the preliminary artillery preparation to begin twelve hours prior to the infantry assault, and it called for the howitzers to pound various target areas within the city limits for the full time period, while certain batteries of 75s registered their guns and then destroyed the barbed-wire obstacles surrounding the town. During the last three hours of the preliminary bombardment, sixteen guns from the 15th FA were to fire nothing but shells of persistent mustard gas (nearly 6,000 in all) into an area just north of Vaux, in an effort to seal off the town from eventual counterattacks from the north.[40]

At 5:00 PM, one hour before the infantry advance, the bombardment was to increase its intensity, with the 155s doubling and then quadrupling their previous rates of fire. All the 75s were to join in at this time as well, firing a combination of high-explosive shell and shrapnel on different target areas (see Map 2.7). At "H-3 minutes," half of the 75s shifted to a standing barrage just in front of the American lines, while the others shifted to a "box" barrage just beyond the final infantry objective. The previous evacuation of the most forward U.S. infantry positions would ensure that the standing barrage would safely come down in front of even the closest German positions. At "H" hour, the rolling barrage was to begin its advance, moving 100 meters in two minutes at first, then in three minutes through the northern outskirts of Vaux.[41]

The plan called for two infantry battalions from the 3rd Brigade, each supported by a dedicated machine-gun and engineer company, to follow the rolling barrage through the town in about thirty minutes, after which they were to consolidate their new line north of Vaux, and then "clean up" the town of any remaining Germans. The latter phase, of clearing

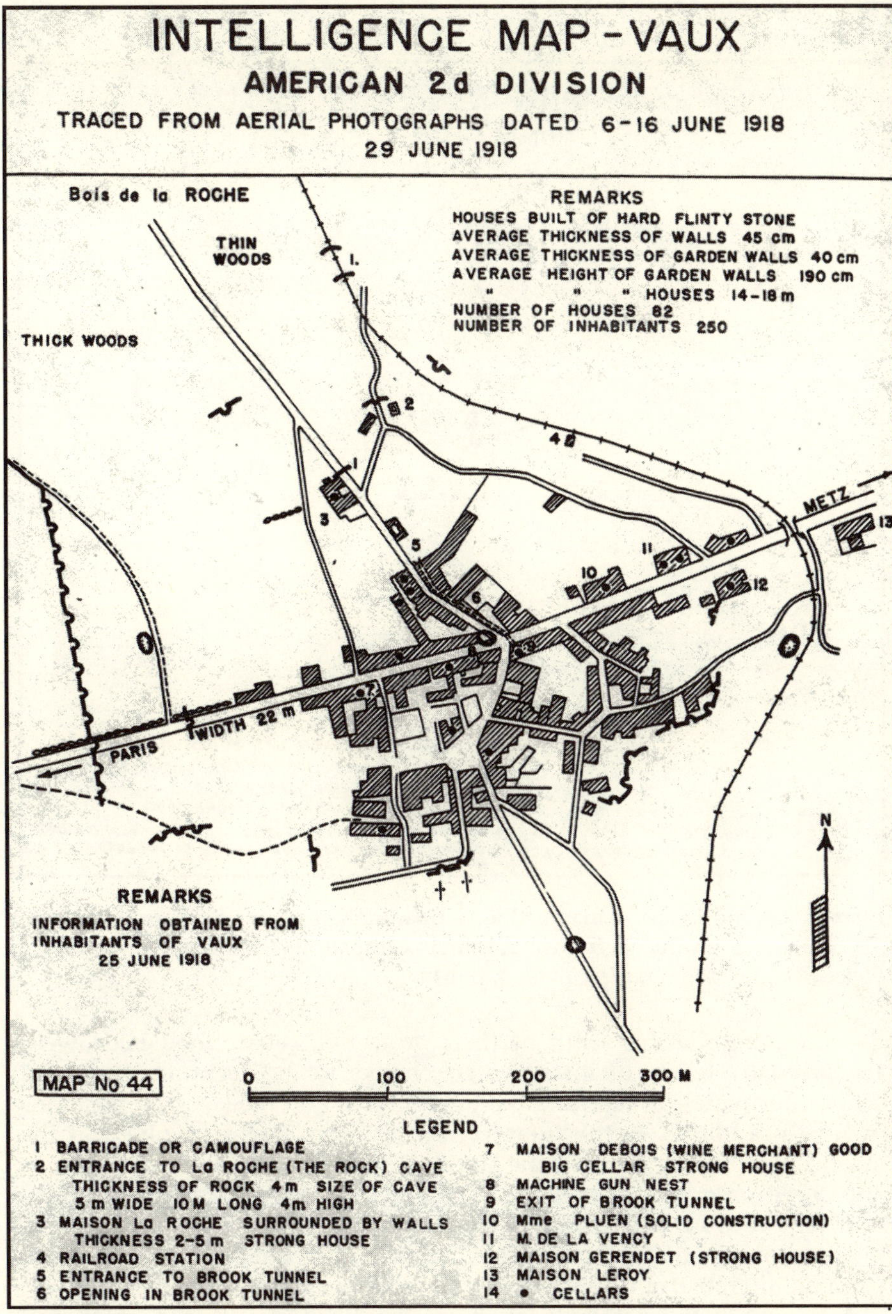

Map 2.5. An AEF intelligence map of Vaux used by the 2nd Division to plan the brilliant set-piece assault of the village. Based on aerial photographs and information from inhabitants, the map detailed the size, location, and strength of each structure in the village. Taken from *United States Army in the World War*, vol. 4, p. 639.

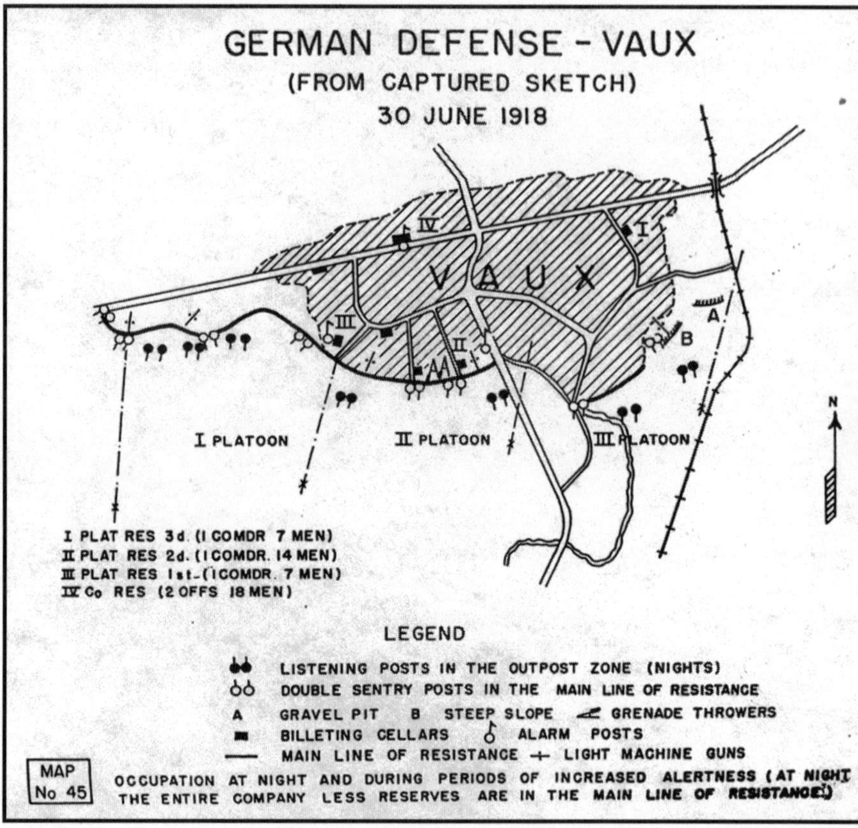

Map 2.6. A 2nd Division map of the German defensive positions around Vaux. The map was produced from a captured German sketch. Taken from *United States Army in the World War*, vol. 4, p. 640.

the surviving Germans out of their protective cellars, had been well-rehearsed by teams of infantrymen, thanks to the accurate maps and other information concerning the town layout and the German positions in it.[42] In addition to the infantry and artillery forces involved in the attack, the division secured the support of five French flying squadrons and three observation balloons. These aerial forces would help adjust artillery fire from both the divisional and the heavy corps artillery, provide reconnaissance and supervision of the battlefield sector, and prevent German aircraft from doing the same for their ground forces.[43]

A review of the detailed nature of the attack plan says a great deal about the different approach taken by the division in its limited, set-piece, artillery-centered attack on Vaux. While the attacks on Belleau Wood were essentially limited in nature, with the exception of the final attack on 25 June, they were never planned with anything close to the

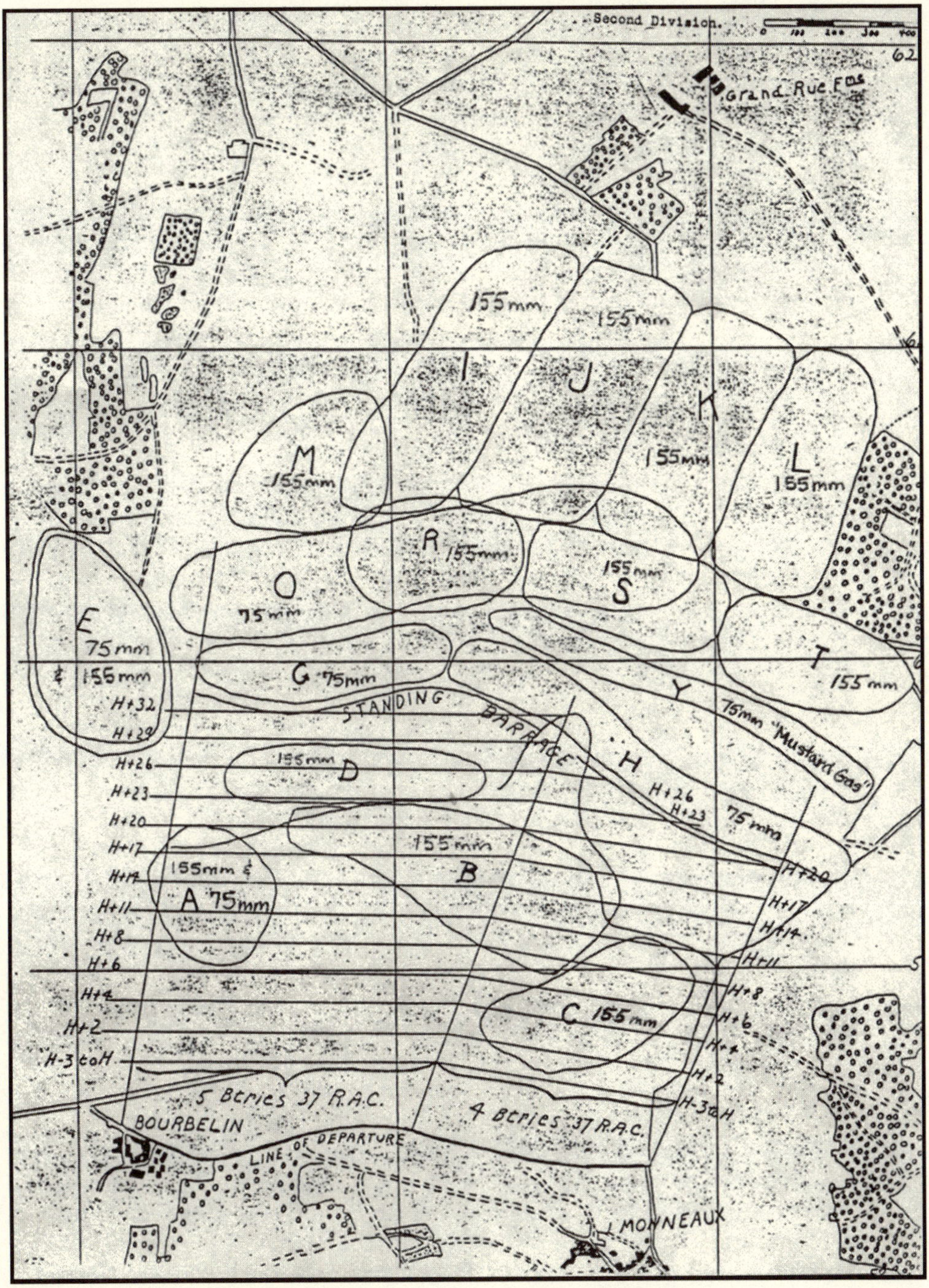

Map 2.7. Barrage chart for the 2nd Division's 1 July 1918 attack on Vaux. Taken from *Records of the Second Division*, vol. 10.

level of precision, or with the emphasis on firepower, that was demonstrated in the Vaux attack plan. The end result of the Vaux operation was also markedly different from the invariably costly, and sometimes ineffective, assaults in Belleau Wood.

The 1 July attack on Vaux, executed by units from the 3rd Brigade, was as clear a victory as it could have been. Brigadier General Edward M. Lewis, the brigade commander, reported that "the plan of action described in Field Order No. 9 . . . was carried out to the letter." The documentary evidence clearly supports the commander's claim. The infantry assault began on time, at 6:00 PM, and within ninety minutes both attacking battalions had reached and secured their objectives. By late evening the plan to "clean up" the town of Vaux was executed just as it had been rehearsed before the battle, and the Germans caught in their protective cellars were taken to the rear.[44]

Early the next morning, the Germans attempted to make a strong counterattack, supported by an artillery bombardment of high-explosive and gas shells. Thanks to the detailed plans made before the 2nd Division's initial attack, the new American positions were so strongly developed, and the plans for defensive fire support so effectively arranged, that the division easily repulsed the counterattack and inflicted numerous casualties on the attackers.[45] Though the opposing artillery forces continued to shell each other's positions for days, this event signaled the end of the operation. The 3rd Brigade held this new line without any threat of withdrawal until the 2nd Division left the front-line trenches over a week later.

After the operation was over, Colonel Albert J. Bowley, who had replaced General Chamberlaine as commander of the 2nd FA Brigade, released a memorandum in which he praised his gunners for their support of the attack. He claimed that their work was "thorough in every particular," that "the strong points of Vaux and the surrounding woods were completely destroyed," and that "the barrage was perfect." It was high praise for a brigade that had been in active operations for less than a month, but everything he said appeared to be correct.[46]

In fact, the infantrymen were even more lavish in their praise for the artillery support. Major A. E. Burton, who commanded the 2nd Battalion of the 9th Infantry Regiment in the initial attack, reported that "the artillery preparation was most thorough and Vaux was a wreck." He stated that "the destructive fire of the artillery . . . was most excellent and far beyond our greatest expectations." Major Burton also claimed that the rolling barrage was "excellently executed."[47] Major Charles E. Elliot, who was the commander of the other attacking infantry battalion, the 3rd Battalion of the 23rd Infantry Regiment, concurred with this assessment, reporting that the rolling barrage was "perfect." Finally, General Lewis stated that "the artillery preparation . . . was most thorough and complete and contributed in large measure to the success of the operations." He concluded by asserting that he could not "refrain from expressing the appreciation of all concerned of the foresight involved in and the admirable preliminary working out of all details, and of the excellence of the artillery work, both in the preparation and during and after the attack [sic]."[48]

No doubt there were many infantrymen in the 3rd Brigade who, after watching the Marines struggle so terribly and for so long in Belleau Wood, joined General Lewis in his appreciation for the superb planning and excellent artillery work that made the attack such a stunning success. Certainly, there were many more infantrymen left alive to be appreciative than most divisional observers would have expected after such an attack. The division suffered only forty-seven deaths during the entire operation, and counting all those killed, wounded, and missing, there were just 328 total casualties. Not an insignificant number, but it measured up well when compared with the terrible losses of the enemy. The German Army had 254 killed, 162 wounded, and 510 missing soldiers to show for the two days of fighting, a total of 926 casualties.[49] Considering that the Americans absorbed their casualties while gaining every assigned objective, and that the German forces suffered theirs while being thrown out of their established positions, the full measure of the American success becomes even more apparent.

The seizing of Vaux was the last operation undertaken by the division before it was finally relieved from front line duty and sent to a rear area for rest on 9 July. The division had effectively been in line since 1 June, and was subjected to relatively heavy artillery fire, including gas shelling on all but four of the days, for its entire tour in the line. From 1 June to 9 July, the 2nd Division suffered approximately 9,700 total casualties, including all killed, wounded, and missing.[50] Many, if not the majority of these casualties, were the result of the intensity and regularity of the artillery activity during the division's extremely long tour in an exceptionally active sector. It would be a mistake to think that poorly orchestrated infantry attacks against well-defended positions caused all or even most of the casualties. Nevertheless, there were clearly too many of the latter, especially in the area of Belleau Wood.[51]

Many factors beyond the control of the 4th Brigade, or even the 2nd Division, worked against the Marines, especially in their earlier operations near the wood. When the French corps commander placed the division in the front line, and quickly ordered it to advance its positions, neither the 4th Brigade nor the division possessed accurate maps with which to plan or carry out the attacks. This was a major problem, especially when the infantry commanders attempted to coordinate artillery support with Marines that were already in the wood. Without accurate maps, and with direct visual observation of artillery fire restricted by the surrounding terrain as well as the wood, it was impossible to deliver safe and accurate fire over the heads of the most forward Marines and into the German positions, which were just meters ahead of them.[52]

Second, poor maps also contributed to the physical and spatial disorientation experienced by infantry commanders fighting in the wood, which further complicated the effort to provide artillery support. General Harbord admonished his subordinate commanders to provide the most detailed and accurate descriptions possible of their positions, since "artillery cannot be safely called for unless infantry positions are accurately known." Unfortunately, few commanders in the wood knew exactly where they were. One method of coping with this problem was to withdraw the infantry all the way back to a clearly defined line, a tactic used with some success in those attacks that utilized strong artillery support.[53]

A third factor that led to the heavy casualties was the lack of sufficient gas shell to carry out what many in the division knew would be the most effective method of eradicating the German positions in the wood—saturate it with persistent gas. While there is no hint that Harbord considered the use of gas to drive out the Germans, other senior officers in the division, such as the artillery commander, Chamberlaine, and Colonel Preston Brown, General Bundy's chief of staff, lamented that the French had not given the division the gas shells it needed to properly saturate the wood.[54]

A fourth external factor that contributed to the huge number of casualties, as well as to the difficulty of providing accurate and effective artillery fire, was the almost total absence of Allied air power in the sector throughout June. Aircraft played an important role in feeding senior commanders accurate information about their forward infantry lines, and an even more critical role in helping to accurately adjust artillery fire directly onto targets. In early June a frustrated General Harbord exclaimed to division headquarters that "the hum of German airplanes is almost constant over our front lines," recommending that the French "show some of that superiority which is referred to in almost every French and British paper recently." The lack of Allied air cover was a double problem, as it meant that the Germans ruled the sky and reaped every reward that came along with it, such as better intelligence and position information, and much more accurate artillery fire.[55]

However, despite these external factors that led to more casualties and mitigated against the effective use of artillery, it is clear that the division, and particularly Harbord's 4th Brigade, displayed a definite lack of appreciation for the firepower placed at its disposal. The Marines made multiple attacks during the three days of 6, 7, and 8 June; each was attempted with inadequate artillery support, and each led to little success and high casualties. After effectively utilizing the supporting artillery for the attacks on 10, 11, and 12 June, Harbord seemed to revert back to his initial methods, counting on the "self-reliant" infantrymen to drive the last machine-gun nests out of the woods. Those attacks, whether attempted by the 7th Infantry Regiment (on 20 June) or his own Marines (23 June), resulted in excessive casualties with little or no advancement to show for them. It was only when Harbord returned to the artillery-centered approach, used in the attack on 25 June, that the 2nd Division cleared Belleau Wood of the last German defenders.

The different methods utilized in the division's attacks on Belleau Wood and Vaux are clear. The division and brigade leadership sent the Marines forward into Belleau Wood without sufficient planning and with inadequate or even nonexistent artillery support. They suffered terribly for it. However, when the division leadership made adequate preparations before the attack and employed overwhelming artillery support, as they did from 10 to 12 June and on the final attack of 25 June, the division proved that it could advance, with acceptable casualty rates, even in an environment that was extremely favorable to the defenders. The attack on Vaux was simply the ultimate demonstration of the division's ability to successfully plan and execute limited, set-piece attacks, making excellent use of those "trench-oriented" skills that had been absorbed during its training period with the French. At Vaux, the officers in the 2nd Division showed that they had learned first-hand at Belleau Wood what the French training officers had warned them about: effective artillery support was crucial to any infantry attack. Pershing's abstract "infantry-based" doctrine had come face to face with the reality of the Western Front, and the reality appeared to win the first battle.

When the 2nd Division came out of the line on 9 July, it went out bloodied, but victorious. Though one can look back with hindsight and see that the German offensive had just about run out of steam by the time the 2nd Division took over its sector, it looked very different to those in France that June 1918.[56] General Tasker H. Bliss, the sole American representative on the Allied Supreme War Council at Versailles, wrote to Secretary of War Newton D. Baker that the "conduct of the Second Division was superb." He claimed that "the French have been wildly enthusiastic about it, and deservedly so. . . . General Foch has expressed his admiration—he was convinced it 'practically saved Paris.' "[57] At the same time, General Harbord commented on his brigade's role in lifting Allied morale by noting that "the effect on the French has been many times out of all proportion to the size of our brigade or the front on which it has operated." An officer on General Pétain's staff referred to the division's timely arrival on the retreating front as "a wonderful transfusion of blood. . . . Life was coming in floods to reanimate the dying body of France." In honor of the "brilliant conduct" of the 4th Brigade during the operations, General Degoutte issued an order proclaiming that Belleau Wood would henceforth be renamed the "Bois de la Brigade de Marine." By all accounts, the 2nd Division had an impressive moral impact on the French Army.[58]

Possibly of even greater significance was the impact the division had on the German Army. The German High Command had pinned its hopes on the supposition that the United States would not be able to get substantial land forces to France in time to save the Allies, and that those Americans who did make it to the Western Front would be inferior soldiers. After the 2nd Division ground up four German divisions during its thirty-five-day tour at the front, the German Army had to readjust its opinions of American fighting capability. An intelligence report issued by the German corps that opposed the 2nd Division stated, "the American 2nd Division may be considered as a very good division, if not even as an attack [i.e., "stormtroop"] unit." Attacks "were carried out with vigor and regardless of losses. The moral effect of our firearms did not materially check the advance of the infantry. The nerves of the Americans are still unshaken . . . the personnel may be considered excellent."[59] After the 2nd Division's operations around Chateau-Thierry, the Germans had to consider that future American divisions would be a force to be reckoned with, and since American troops were arriving in France by the hundred thousands every month, it spelled doom for German hopes of victory.

NOTES

1. "Summary of Battle of Chateau-Thierry, May 31st to July 10th, 1918," Major General John A. Lejeune, in U.S. Army, 2nd Division, *Records of the Second Division (Regular)*, (Washington, D.C.: Army War College, 1924) (hereafter *RSD*), vol. 6.

2. Edward Coffman, *The War to End All Wars* (New York: Oxford University Press, 1968), p. 213; Smythe, *Pershing*, p. 131.

3. *USAWW*, 4: 88; General Operations Order, HQ XXI Corps, 1 June 1918, Folder 445, Entry 318-30.1, RG 165, NA; Lieutenant William D. Bickham and Major William E. Burr, "Second Field Artillery Brigade History," pp. 8–9, undated typescript, in 2nd FA Brigade Historical File, Entry 1246-11.4, RG 120, NA; Captain John D. Clark, "My War," p. 9, unpublished memoir, 15th FA folders, 2nd Division boxes, World War I Manuscript Collection at USAMHI.

4. "Report of Second Division Movements from May 31st to July 10th, 1918," HQ 2nd FA Brigade, 19 December, 1918, *RSD*, vol. 7; "Diary Containing Field Orders and Instruction, 4th Brigade, M.C. (2d Division), May 30–June 30, 1918," p. 4, Folder 79, 2nd Division files, G-3 Reports, Entry 270, RG 120, NA.

5. While the light gun regiments were subject to the calls of the infantry brigades, control of the howitzer regiment, the 17th FA, was maintained by the brigade commander. Field Order No. 1, HQ 2d FA Brigade, 3 June 1918, *RSD*, vol. 9; Captain John D. Clark, "My War," p. 9, unpublished memoir, 15th FA folders, 2nd Division boxes, World War I Manuscript Collection at USAMHI.

6. The French troops that were relieved by the 2nd Division had been in line for six days and had fallen back each day that they were in line. When the 2nd Division set up its line of resistance, it was told that "no support should be expected under any pretext whatever." General Operations Order, HQ XXI Corps, 1 June 1918, Folder 445, Entry 318-30.1, RG 120, NA; James G. Harbord, *Leaves from a War Diary* (New York: Dodd, Mead, 1925), p. 291; "Report of Second Division Movements from May 31st to July 10th, 1918," HQ 2nd FA Brigade, 19 December, 1918, *RSD*, vol. 9.

7. "Diary Containing Field Orders and Instruction, 4th Brigade, M.C. (2d Division), May 30–June 30, 1918," pp. 4–5, Folder 79, 2nd Division files, G-3 Reports, Entry 270, RG 120, NA; USAWW, 4: 161.

8. General Operations Order, XXI Corps, 5 June 1918, in *USAWW*, 4: 143.

9. Smythe, *Pershing*, p. 139; ABMC, *American Armies*, pp. 46, 49; Coffman, *War to End All Wars*, p. 221.

10. Harbord, *American Army*, pp. 296, 300.

11. "Journal of Activities of the 2nd Division from May 31st to June 30th," Folder 13, 2nd Division files, G-3 Reports, Entry 270, RG 120, NA.

12. Field Order No. 1, 4th Brigade, in "Diary Containing Field Orders and Instruction, 4th Brigade, M.C. (2d Division), May 30–June 30, 1918," p. 7, Folder 79, 2nd Division files, G-3 Reports, Entry 270, RG 120, NA.

13. By this date, the 2nd FA Brigade was augmented by another French light gun regiment and two French battalions of 155s, giving the division a total of 167 guns. The divisional front had been shortened to under twelve kilometers, giving the division a gun density of one gun for each sixty-seven meters of front. "Report of Operations, 2d Field Artillery Brigade, 2d Division, During Month of June, 1918," p. 16, by Brigadier General William Chamberlaine, submitted to the Adjutant General of the Army, 10 September 1919, 2nd FA Brigade Historical File, Entry 1246-33.6, RG 120, NA; Orders No. 1, HQ 2d FA Brigade, 5 June 1918, *RSD*, vol. 9.

14. "Diary Containing Field Orders and Instruction, 4th Brigade, M.C. (2d Division), May 30–June 30, 1918," p. 8, Folder 79, 2nd Division files, G-3 Reports, Entry 270, RG 120, NA.

15. General Operations Order, XXI Corps, 5 June 1918, in *USAWW*, 4: 143; Field Order No. 2, 4th Brigade, in "Diary Containing Field Orders and Instruction, 4th Brigade, M.C. (2d Division), May 30–June 30, 1918," pp. 9–10, Folder 79, 2nd Division files, G-3 Reports, Entry 270, RG 120, NA.

16. Field Order No. 3, HQ 2d FA Brigade, 6 June, 2:00PM, in *USAWW*, 4: 363.

17. Harbord, *American Army*, pp. 289–92.

18. "Diary Containing Field Orders and Instruction, 4th Brigade, M.C. (2d Division), May 30–June 30, 1918," pp. 11–14, Folder 79, 2nd Division files, G-3 Reports, Entry 270, RG 120, NA.

19. "Diary Containing Field Orders and Instruction, 4th Brigade, M.C. (2d Division), May 30–June 30, 1918," pp. 13–16, Folder 79, 2nd Division files, G-3 Reports, Entry 270, RG 120, NA.

20. It would be the costliest day in Marine Corps history until the attack on Tarawa atoll in 1943. Coffman, *War to End All Wars*, p. 217.

21. "Diary Containing Field Orders and Instruction, 4th Brigade, M.C. (2d Division), May 30-June 30, 1918," p. 22, Folder 79, 2nd Division files, G-3 Reports,

Entry 270, RG 120, NA; "Summary of Battle of Chateau-Thierry, May 31st to July 10th, 1918," Major General John A. Lejeune, *RSD*, vol. 6.

22. Order, XXI Corps, 7 June 1918, in *USAWW*, 4: 403.

23. "Report of Operations, 2d Field Artillery Brigade, 2d Division, During Month of June, 1918," p. 23, by Brigadier General William Chamberlaine, submitted to the Adjutant General of the Army, 10 September 1919, 2nd FA Brigade Files, Entry 1246-33.6, RG 120, NA; "Diary Containing Field Orders and Instruction, 4th Brigade, M.C. (2d Division), May 30–June 30, 1918," pp. 23–25, Folder 79, 2nd Division files, G-3 Reports, Entry 270, RG 120, NA.

24. "Report of Operations, 2d Field Artillery Brigade, 2d Division, During Month of June, 1918," p. 25, by Brigadier General William Chamberlaine, submitted to the Adjutant General of the Army, 10 September 1919, 2nd FA Brigade Files, Entry 1246-33.6, RG 120, NA; "Diary Containing Field Orders and Instruction, 4th Brigade, M.C. (2d Division), May 30–June 30, 1918," pp. 28–29, Folder 79, 2nd Division files, G-3 Reports, Entry 270, RG 120, NA.

25. "Report of Operations, 2d Field Artillery Brigade, 2d Division, During Month of June, 1918," p. 25, by Brigadier General William Chamberlaine, submitted to the Adjutant General of the Army, 10 September 1919, 2nd FA Brigade Historical File, Entry 1246-33.6, RG 120, NA.

26. "Diary Containing Field Orders and Instruction, 4th Brigade, M.C. (2d Division), May 30–June 30, 1918," pp. 29–33, Folder 79, 2nd Division files, G-3 Reports, Entry 270, RG 120, NA.

27. "Diary Containing Field Orders and Instruction, 4th Brigade, M.C. (2d Division), May 30–June 30, 1918," pp. 39–40, Folder 79, 2nd Division files, G-3 Reports, Entry 270, RG 120, NA.

28. The division first thought that this attack was "a complete success in every respect" and that they had "captured the entire woods." This proved to be incorrect, as the operations of the next two weeks showed. "Report of Operations" for the period ending 8:00 PM, June 11, *RSD*, vol. 6; Spaulding and Wright, *The Second Division*, p. 75; "Diary Containing Field Orders and Instruction, 4th Brigade, M.C. (2d Division), May 30–June 30, 1918," pp. 34–37, Folder 79, 2nd Division files, G-3 Reports, Entry 270, RG 120, NA.

29. "Diary Containing Field Orders and Instruction, 4th Brigade, M.C. (2d Division), May 30–June 30, 1918," p. 39, Folder 79, 2nd Division files, G-3 Reports, Entry 270, RG 120, NA.

30. "Diary Containing Field Orders and Instruction, 4th Brigade, M.C. (2d Division), May 30–June 30, 1918," p. 40, Folder 79, 2nd Division files, G-3 Reports, Entry 270, RG 120, NA.

31. Journal of Operations, 12 June 1918, *RSD*, vol. 6; "Diary Containing Field Orders and Instruction, 4th Brigade, M.C. (2d Division), May 30–June 30, 1918," pp. 40–41, Folder 79, 2nd Division files, G-3 Reports, Entry 270, RG 120, NA.

32. "Summary of Battle of Chateau-Thierry, May 31st to July 10th, 1918," Major General John A. Lejeune, *RSD*, vol. 6.

33. "Diary Containing Field Orders and Instruction, 4th Brigade, M.C. (2d Division), May 30–June 30, 1918," pp. 48–49, Folder 79, 2nd Division files, G-3 Reports, Entry 270, RG 120, NA.

34. "Diary Containing Field Orders and Instruction, 4th Brigade, M.C. (2d Di-

vision), May 30–June 30, 1918," pp. 54–55, Folder 79, 2nd Division files, G-3 Reports, Entry 270, RG 120, NA.

35. "Diary Containing Field Orders and Instruction, 4th Brigade, M.C. (2d Division), May 30–June 30, 1918," p. 59, Folder 79, 2nd Division files, G-3 Reports, Entry 270, RG 120, NA.

36. "Diary Containing Field Orders and Instruction, 4th Brigade, M.C. (2d Division), May 30–June 30, 1918," p. 60, Folder 79, 2nd Division files, G-3 Reports, Entry 270, RG 120, NA; "Report of Operations," 25 June 1918, *RSD*, vol. 6.

37. The defending Germans reported 430 total casualties. Spaulding and Wright, *The Second Division*, p. 69; "Diary Containing Field Orders and Instruction, 4th Brigade, M.C. (2d Division), May 30–June 30, 1918," p. 64, Folder 79, 2nd Division files, G-3 Reports, Entry 270, RG 120, NA; Journal of Operations, 26 June 1918, *RSD*, vol. 6.

38. "Summary of Battle of Chateau-Thierry, May 31st to July 10th, 1918," Major General John A. Lejeune, *RSD*, vol. 6.

39. See "Information Bulletin—Vaux," 2nd Division, 29 June 1918, and associated maps in *USAWW*, 4: 636–40.

40. Field Orders No. 9, HQ 2d Division, in *USAWW*, 4: 645–46.

41. Field Orders No. 9, HQ 2d Division, in *USAWW*, 4: 645–46; Spaulding and Wright, *The Second Division*, pp. 74–75.

42. Field Orders No. 9, HQ 2d Division, in *USAWW*, 4: 645–46; Spaulding and Wright, *The Second Division*, pp. 74–75; Special Operations Report, 2d Battalion, 9th Infantry, 31 July 1918, in *USAWW*, 4: 668.

43. Field Orders No. 9, HQ 2d Division, in *USAWW*, 4: 645–46.

44. Special Operations Report, 3rd Brigade, 2 July 1918, in *USAWW*, 4: 674; "Report of 2d Battalion, 9th Infantry, Taking of Vaux at 6pm," 11:30 PM, 1 July 1918, in Folder 57, 2nd Division files, G-3 Reports, Entry 270, RG 120, NA.

45. During the German counterattack, the 3rd Brigade took an additional 180 prisoners. Special Operations Report, 3rd Brigade, 2 July 1918, in *USAWW*, 4: 674.

46. Memorandum, 2d FA Brigade, 2 July 1918, in *USAWW*, 4: 681. Special Operations Report, 2d Battalion, 9th Infantry, 3 July 1918, in *USAWW*, 4: 668.

47. "Report of 2d Battalion, 9th Infantry, Taking of Vaux at 6 pm," 11:30 PM, 1 July 1918, in Folder 57, 2nd Division files, G-3 Reports, Entry 270, RG 120, NA; Special Operations Report, 2d Battalion, 9th Infantry, 3 July 1918, in *USAWW*, 4: 668.

48. Special Operations Report, 3rd Brigade, 2 July 1918, in *USAWW*, 4: 674.

49. Journal of Operations, 2 July 1918, *RSD*, vol. 6; Spaulding and Wright, *The Second Division*, p. 78.

50. This figure must include those "slightly wounded and returned to duty," as well as those twice wounded, and is taken from the report of the 2nd Division Commander (Major General John A. Lejeune) to Assistant Chief of Staff, G-3, GHQ, AEF, Subject: "Operations Information," 30 December 1918, in Reports and Other Records Regarding the History of AEF Units, Divisional Skeleton Histories, Entry 443, RG 120, NA.

51. On one night alone, 13/14 June, a surprise gas bombardment of a rear area caused nearly 800 gas casualties. The division regularly suffered well over a hundred casualties a day, even when not making any attacks.

52. Lieutenant William D. Bickham and Major William E. Burr, "Second Field Artillery Brigade History," typescript, p. 11, in 2nd FA Brigade File, Entry 1246-11.4, RG 120, NA.

53. "Diary Containing Field Orders and Instruction, 4th Brigade, M.C. (2d Division), May 30–June 30, 1918," p. 25, Folder 79, 2nd Division files, G-3 Reports, Entry 270, RG 120, NA.

54. Rexmond C. Cochrane, *Gas Warfare at Belleau Wood, June 1918*. Study #1, U.S. Army Chemical Corps Historical Studies: Gas Warfare in World War I (Army Chemical Center, Maryland, 1957), p. 71; "Report of Operations, 2d Field Artillery Brigade, 2d Division, During Month of June, 1918," p. 29, by Brigadier General William Chamberlaine, submitted to the Adjutant General of the Army, 10 September 1919, 2nd FA Brigade Files, Entry 1246-33.6, RG 120, NA.

55. The 16 June entry in the divisional Journal of Operations stated "the activity of the enemy air service during the past few days has been very marked. It maintains practically a continuous reconnaissance over our lines, particularly over our forward zone. At times they come in large patrols. 57 plans flew over during the day. . . . Our air service, which is entirely French at present, is much less active," *RSD*, vol. 6; "Diary Containing Field Orders and Instruction, 4th Brigade, M.C. (2d Division), May 30–June 30, 1918," pp. 5–6, Folder 79, 2nd Division files, G-3 Reports, Entry 270, RG 120, NA.

56. At least one French Army study on the German Chemin des Dames Offensive considered the "Battle of the Aisne," as they called it, over on 4 June, the day the 2nd Division took over the front. See GHQ of the Armies of the North and Northeast (French), "The Battle of the Aisne (27 May–5 June)," 10 July 1918, in French Military Records Relating to World War I, 1917–1919, Entry 403—20.2-20.9, Records of the War Department General and Special Staffs, RG 165, NA.

57. Letter, General Bliss to Secretary Baker, 14 June 1918, in Harbord, *The American Army*, p. 299.

58. Harbord, *Leaves*, p. 294; Smythe, *Pershing*, p. 138; Spaulding and Wright, *The Second Division*, pp. 94.

59. "Intelligence Report," IV Reserve Corps, 17 June 1918, in *USAWW*, 4: 607.

3

Soissons: A Brilliant Advance, at a Terrible Price

. . . the turning point of the world war.[1]

General Robert L. Bullard
Commanding General, American Second Army

Though the troops in the 2nd Division hoped that AEF GHQ was going to grant them a lengthy period of rest to absorb the thousands of replacements that had joined the division in the past month, the Allied High Command had other plans. On 14 July, just four days after pulling its guns out of their forward positions near Belleau Wood and Vaux, the 2nd FA Brigade received orders from the French Tenth Army to detach itself from the division, move twenty kilometers north to the town of Betz, and await further orders. General Bowley gave instructions for the movement to begin that night.

The following day, General Bundy left the division to command the American VI Corps. General Harbord assumed command of the division. That very night Harbord received orders to move the remainder of his division north to the area under the control of General Charles Mangin, the commander of the French Tenth Army. The movement started on 16 July, with some units traveling by foot, others by truck, but all under the control of the French transportation system. Though no one in the division knew it, not even the divisional commander, the 2nd Division was on its way to take part in a massive offensive scheduled to begin on the early morning of 18 July.[2]

When Harbord reported to Mangin's headquarters on the evening of 16 July, he found that the army commander had already left for the Tenth

Army's advanced command post, and nobody else was able to give him any information about his division's future mission. Harbord summed up his thoughts concerning the situation by stating, "a division of twenty-eight thousand men, the size of a European army corps, had been completely removed from the control of its responsible commander, and deflected by truck and by marching, through France to a destination unknown to any of the authorities responsible either for its supply, its safety, or its efficiency in the coming attack. . . . This within thirty hours of a decisive battle."[3]

Eventually Harbord found Bowley, who introduced him to General Pierre E. Berdoulat, the commander of the French XX Corps, under whose guidance the division was to make the upcoming attack. While Harbord looked for assurance that his division would arrive in time to prepare for the attack, as well as for detailed information concerning the upcoming operation, the only help he received from the XX Corps was "their Corps Attack Orders and a liberal supply of maps."[4] At least this put the division a step ahead of its entry into the Chateau-Thierry sector. Immediately Harbord and Colonel Brown, his chief of staff, set about preparing the division attack order "for a division whose location was unknown, whose presence at the appointed time for the attack was by no means certain and which, furthermore, was to attack over terrain that could not be reconnoitered."[5]

By early morning Harbord's staff had mimeographed the order, but little else had improved. Although the artillery brigade was in position behind the attack front, Harbord still had no idea as to the whereabouts of any other unit of his division. Eventually, at various times during the day, the locations of some of the units became known. By mid-day on 17 July, only the 23rd Infantry Regiment was near the attack area. The 6th Marine Regiment, though the farthest away, had been designated as the corps reserve, and was therefore not critical. The other two regiments, the 9th Infantry and 5th Marines, both had to make forced marches that night, with some units even having to double-time the final few kilometers in order to reach the jump-off position on time. The division staff spent the entire day of 17 July "in a desperate effort to gather up scattered remnants, searching for lost platoons and companies, and in locating ammunition."[6] Less than twenty-four hours before the attack, Harbord could not even be sure that all of his attacking units would reach their jump-off positions before the attack started.

While General Harbord had written his division attack order, there were few important decisions left to his discretion. The senior French army and corps commanders dominated this attack more than any other 2nd Division operation of the war. Plans called for the division to attack at 4:35 AM on 18 July on the right flank of the French XX Corps. The XX Corps was composed of the U.S. 1st Division on the left, the famous

1st Moroccan Division (with French Foreign Legion) in the center, and the 2nd Division on the right. The XX Corps, along with two other French corps in Mangin's Tenth Army, was to make a sudden and rapid advance into the northern shoulder of the giant German salient that extended south of the Aisne River some forty kilometers to Chateau-Thierry (see Map 3.1). The stated "final objective" of the French XX Corps was an ambiguous line nearly twenty kilometers behind the German positions. However, the main goal of the attack was to cut the Soissons-to-Chateau-Thierry highway and railroad, which were critical transportation arteries for the German Army. If the Allied advance could cut off that transportation and supply route, the German forces south of the Vesle River would probably have to evacuate that portion of the salient.

The attack plans of the French XX Corps and the 2nd Division directed the division to advance eastward on a front of three kilometers. While the attack order specified three intermediate objectives for the division, for all intents and purposes the plans called for an advance with an unlimited penetration. In this respect, as well as others, the operation did not conform to the conservative pattern of attack that had come to typify Allied attacks during the past year. Not only did Mangin give the attacking divisions unlimited objectives, but in another break with the dominant Allied doctrine, he did not base the attack on the overwhelming use of artillery support in all phases of the operation.

Concerning artillery employment, Mangin's counteroffensive relied on two significant departures from the most recent French attacks. First, like the British attack on Cambrai in November 1917, no preliminary artillery bombardment would precede the infantry assault. Mangin clearly based the attack more on surprise than on massive artillery support. The German defensive positions were relatively new and weak, and Mangin gave each attacking division approximately forty-eight tanks to help drive through the German lines. While there was to be no significant artillery fire before the infantry attack, the infantry and tank advance was to be covered by a rolling barrage that was to be "as dense as possible."

The deep, virtually unlimited objectives of the attack plans necessitated a second deviation from previous rules of artillery employment. Mangin ordered the infantry forces to continue the attack beyond the range of the established artillery support. The rolling barrage could cover the infantry advance for just the first three to four kilometers (i.e., just to the first intermediate objective). After that distance the attacking divisions had to rely on advancing their light field artillery forces and coordinating the fire of those mobile batteries with the most advanced infantry battalions. Effective coordination between the artillery and the infantry was extremely difficult to achieve once the rolling barrage had been outrun, but that is what the plan required.

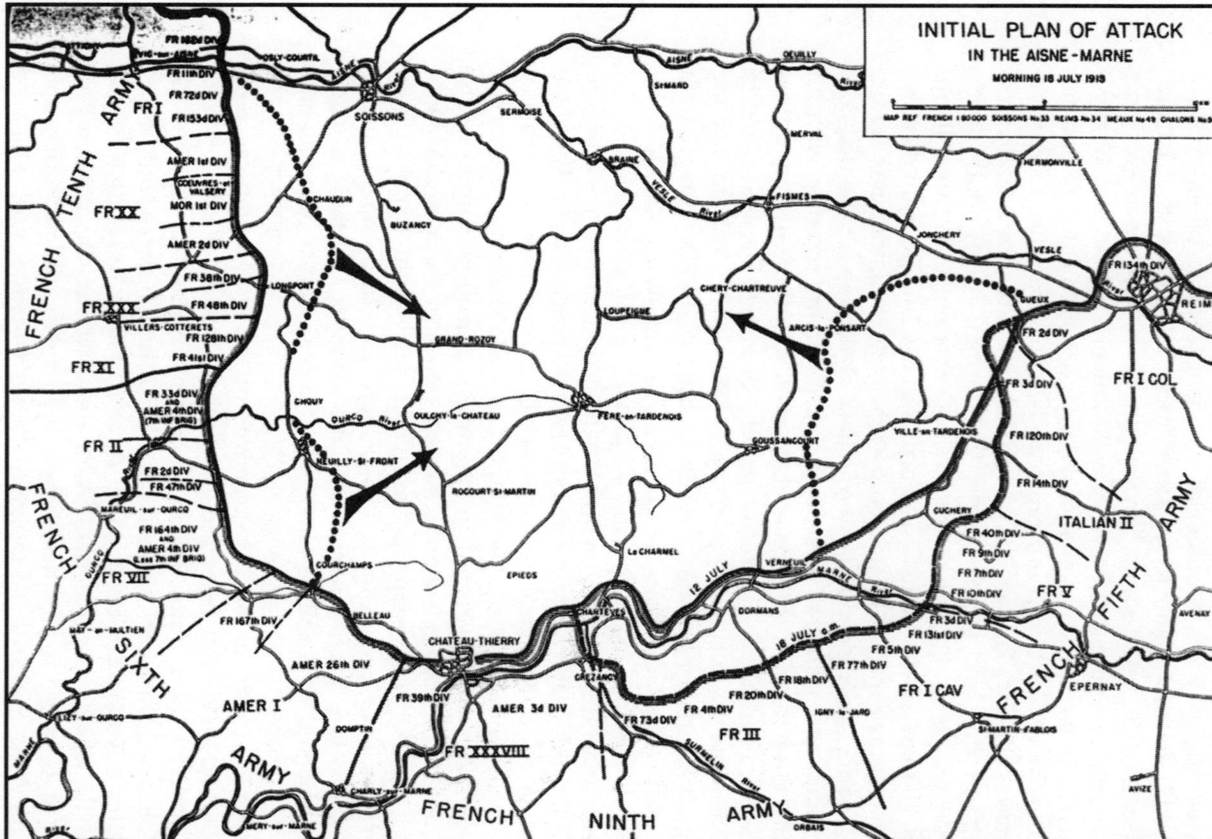

Map 3.1. Initial plan of attack on the Aisne-Marne salient, 18 July 1918. The 2nd Division attacked as part of the French XX Corps in the French Tenth Army in the northwest part of the salient. Taken from *United States Army in the World War*, vol. 5, p. 231.

It might come as no surprise that General Pershing was enthusiastic about this giant counterattack. Though he played no part in planning the operation, he had urged the Allied generalissimo, Marshal Ferdinand Foch, to undertake such an aggressive offensive, and Pershing wanted his AEF divisions to be a part of it.[7] Pershing also would have been excited about the aggressive plan employed by General Mangin, which called for a deep penetrating infantry advance that was not limited by the range of the supporting artillery. If things went according to plan, the French and American forces were to finally engage in "open warfare."

However, despite Pershing's hopes and plans, the American divisions had not received much training in the conduct of mobile, "open warfare," as Chapter 1 demonstrated. Likewise, the previous combat experiences of the two American divisions in Mangin's army, such as the capture of the town of Cantigny by 1st Division and the fighting of the 2nd Division at Belleau Wood and Vaux, did not prepare them for the type of operation in which they would soon take part. In fact, while the 2nd Division fought courageously, gained a great tactical success, and played a major role in the strategic victory achieved by the offensive, it sustained extremely heavy casualties during its advance and demonstrated that it did not yet possess the capability to efficiently fight in an "open warfare" environment.

In his attack order, General Bowley attempted to give the division the guidance it needed to properly coordinate artillery support in the zone beyond the prearranged rolling barrage. After firing the rolling barrage to the first intermediate objective (or "first phase"), each American light gun regiment shifted to provide dedicated support to one of the infantry brigades. Bowley placed the 12th FA "at the disposal" of the 4th Marine Brigade, whose 5th Marine Regiment was to attack on the left half of the division sector. He ordered the 15th FA to support the two infantry regiments of the 3rd Brigade, which would advance on the right. Bowley's plans directed both the 12th and 15th FA Regiments to be prepared to advance immediately after completing the rolling barrage. The 17th FA was to advance only when ordered forward by General Bowley.[8]

Bowley warned his regiments that "after the first phase, the success of the artillery fire will depend exclusively upon . . . the preparations made to move the artillery forward promptly, the selection of positions from which the fire can be controlled nearby by observers, the steps taken to preserve liaison with the higher units," and "the efforts made to keep supplied with ammunition."[9] Strangely, the plans made no mention of the critical importance of maintaining effective liaison between the light gun regiments and the forward infantry battalions, which were to have the greatest and most urgent need of effective artillery support during the later stages of the attack.

The French XX Corps augmented the 2nd FA Brigade with one thirty-six-gun French regiment of 75s. The 2nd FA Brigade received no additional 155mm howitzers, which meant that the division attacked with just 108 total guns in support, spread out across its three kilometer front.[10] Since only the 75s fired the rolling barrage, the gun density for the barrage was approximately one gun for each thirty-five meters of front. This support only continued through the first objective of the attack. The howitzers were required to fire in certain zones in advance of the barrage, but since the division received no augmentation in the area of heavy artillery, there was only one heavy artillery piece for every 125 meters of front.[11] The relative scarcity of artillery support would not cause the surprise attack to fail, but it contributed to excessive casualties.

At 4:35 AM, the guns of the French XX Corps opened up on the German defensive positions. Incredibly, all the American infantry units reached their jump-off positions on time, although some had to run the final few kilometers to reach their attack positions in time to continue forward behind the advancing barrage. However, while the infantry units made it in time, other important parts of the division did not (see Map 3.2). The traffic conditions in the rear of the XX Corps area were so "indescribably bad" that the divisional supply of machine guns, 37mm guns, hand grenades, and trench mortars were all stuck in the massive traffic jams that led to the front line. The infantrymen in the 2nd Division "jumped off" with nothing but their rifles, bayonets, and a meager supply of personal ammunition.[12] On the front of the 9th Infantry Regiment, even the French tanks were late, and Colonel Leroy Upton ordered his regiment to advance without them rather than lose contact with the rolling barrage.[13]

Despite these pronounced disadvantages, the attacking infantry battalions surged forward and caught the defending Germans completely unprepared. After advancing between two and three kilometers within an hour, the leading waves reached their first intermediate objective across the entire division line. Company commanders reported weak resistance from the surprised enemy, and the casualties were not heavy. After a twenty-minute halt behind the standing barrage put down beyond that first objective, the leading battalions continued the attack, this time without any semblance of a rolling barrage to cover their advance. Between 7:00 and 8:00 AM, the light guns began to displace forward, meaning that the only artillery fire available was from the howitzers, which, for safety reasons, lifted their fire all the way to the second intermediate objective (see Map 3.3). The infantry was on its own for the next two kilometers, and the casualties began to mount.[14]

The 23rd Infantry Regiment, on the far right, made the best time, reaching the second objective by 7:00 AM, and the third (or normal) objective by 9:30 AM. The advance was a bit slower in the center and

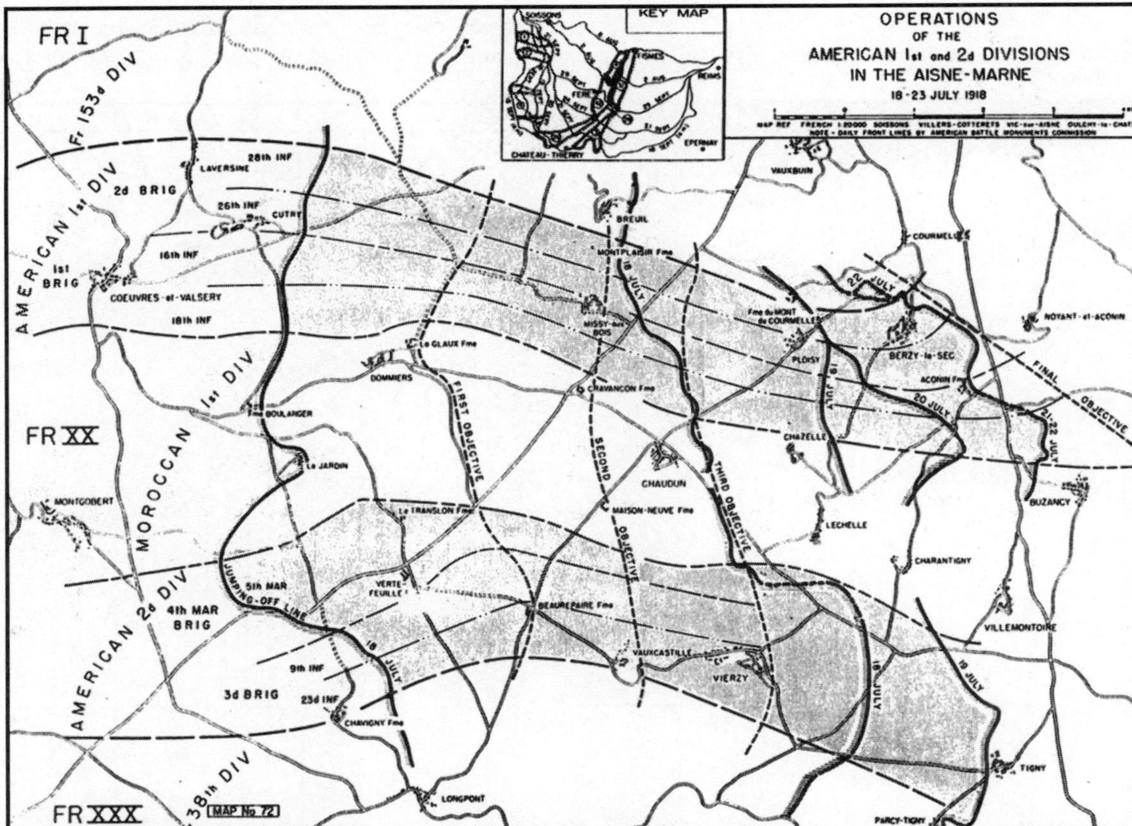

Map 3.2. 2nd Division operations in the Aisne-Marne or Soissons offensive, 18–19 July 1918. Notice the exposed position of the 2nd Division on the night of the 18th, after it had advanced much farther than the Moroccan and French divisions on its flanks. *United States Army in the World War*, vol. 5, p. 309.

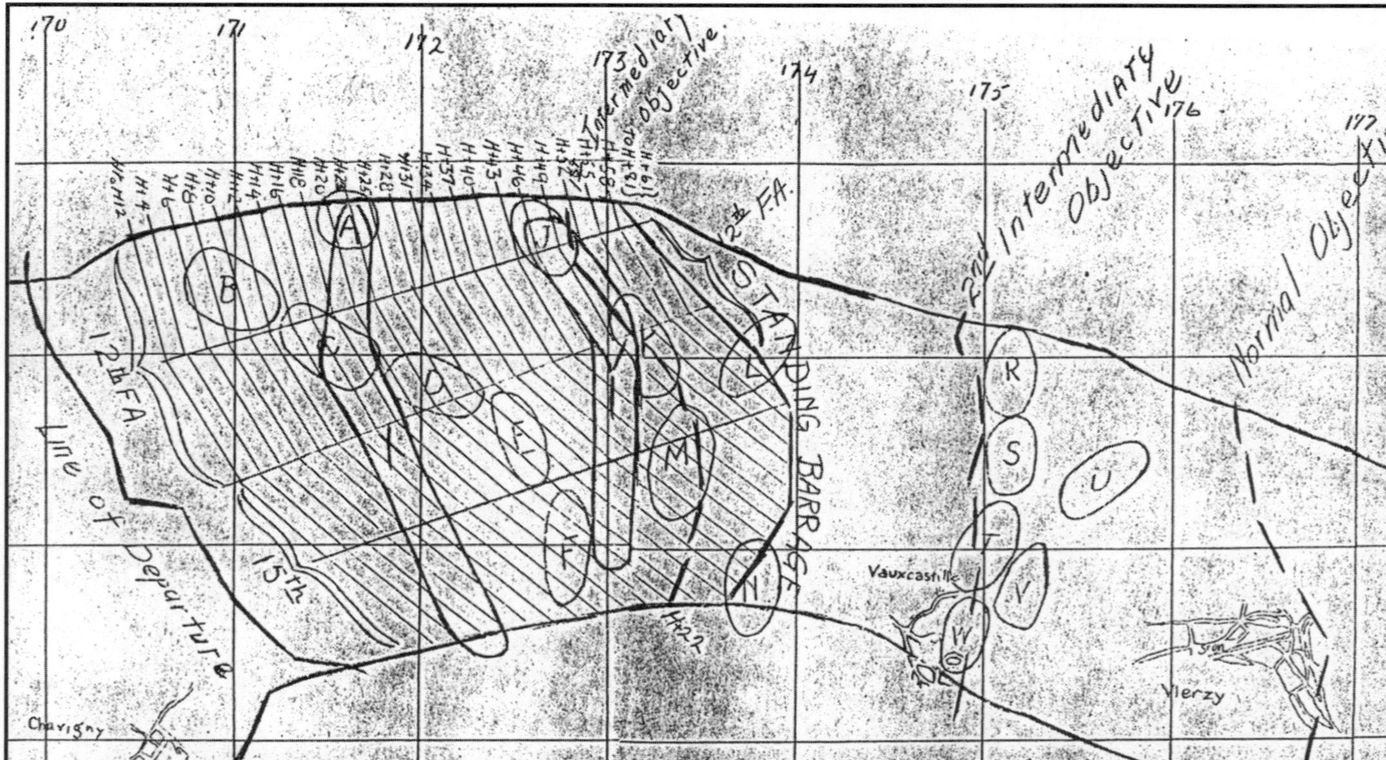

Map 3.3. Barrage chart for the 2nd Division's 18 July attack toward Soissons. The lack of preprogrammed artillery support for the later portion of the attack is evident. *Records of the Second Division*, vol. 10.

the left, where the 5th Marines veered into the sector of the lagging 1st Moroccan Division in order to suppress the galling flank fire coming from the German support lines. In fact, the Marines went far enough north in the hope of securing their left flank that they met up with troops from the U.S. 1st Division, who had extended their right flank for the same reason. Nevertheless, the leading waves of the 2nd Division reached the third objective all along the line by noon. At this point, infantry battalion commanders reported casualties as "quite heavy." Many units were thoroughly disorganized by the attack, as it had required two challenging changes of direction.[15]

While the infantrymen spent the next six hours consolidating their positions and attempting to reorganize themselves, French army and corps commanders prepared for a breakthrough. They rushed cavalry units into the 2nd Division sector and directed the division to continue the assault. At 1:30 PM, General Harbord ordered the attack to begin again immediately. However, due to transportation and liaison problems that slowed transmission of the attack order to the leading infantry units, the attack did not take place until after 7:00 PM. The second attack was apparently attempted without any substantial artillery support. In fact, none of the infantry commanders mention *any* Allied artillery support in their operations reports, unless it was to say that they attacked without it, as was done by Major R. S. Keyser of the 5th Marine Regiment.[16]

The evening attack pushed the American line some two kilometers past the town of Vierzy before stopping due to darkness and the "very severe" casualties sustained at the hands of German machine-gun and artillery fire. The German fire was especially fierce from the north, where the Moroccans, after having nearly caught up to the 2nd Division during the day, failed to advance at all in their evening attack. After its own evening assault, the 2nd Division was done for the day, and nearly for the operation. Its attacks of 18 July, in which all but the first were without any appreciable artillery support, had driven a salient more than two kilometers beyond the French divisions on either side. However, those attacks also decimated the division's three attacking infantry regiments. When Colonel Upton received his availability reports the next morning, he had less than 400 men still organized in his three battalions. Colonel Paul B. Malone reported that the 23rd Infantry Regiment finished the operation with just 37 officers and 1,478 men, which meant that he had lost well over half of his regiment. The casualties in the 5th Marine Regiment were heavy as well.[17]

Despite these casualty figures, the division received corps orders at 2:00 AM on 19 July directing it to continue the attack at 4:00 AM. Unable to attack with such short notice, General Harbord ordered the 6th Marine Regiment to advance from its reserve position and attack at 7:00 AM. He directed an hour-long artillery preparation before the infantry assault. However, by the time the 6th Marine Regiment reached their jump-off position, it was 8:15 AM, and the preliminary artillery fire had long since ceased. Incredibly, despite the lack of coordinated artillery support, the Marines still advanced more than two kilometers, bringing the forward line to within a few hundred meters of the Soissons-to-Chateau-Thierry highway before the attack broke down in the face of strong enemy machine-gun and artillery fire. The 6th Marine Regiment paid an excessively high price in casualties for the two-kilometer advance, suffering approximately 1,300 in killed and wounded from the 2,450 Marines that began the operation.[18]

After the attack on the morning of 19 July, General Berdolat pulled the 2nd Division's infantry regiments out of the line and replaced them with troops from a French colonial division. During the two days of fighting, the division had suffered a total of 3,942 casualties, and all of its infantry units were completely exhausted.[19] However, it had effectively accomplished its mission. The division advanced nearly eleven kilometers into enemy lines, brought a critical transportation and supply artery within close range of artillery fire, and had all but assured that the German Army would have to withdraw its forces in the Marne salient back to the Vesle River.[20]

After the 2nd Division left the front lines, the commander of the 1st Moroccan Division reported on its effectiveness in the offensive. He stated that both American divisions "fought remarkably, with a bravery beyond all praise." He also complimented their willingness to offer assistance when it was needed. However, he added that "they still have a great deal to learn." Specifically, he claimed that they attacked with "too great a density," executed "massed movements under fire," and suffered "unnecessary casualties."[21] He was correct on all accounts.

A review of the operations reports reveals two critical weaknesses with 2nd Division operations in the Aisne-Marne counteroffensive. First, everything was done in a rush. As previously stated, the division's entry into the sector as well as its initial assault both took place in a hurried, almost haphazard fashion. Nearly all of the blame for this rests with the French High Command. The fact that the 2nd Division was able to advance at all on the morning of 18 July is a credit to the officers and noncommissioned officers who pushed their men to the limit to reach the jump-off position in time to attack. The division also had very little say in how it was to attack or employ its artillery on the first day. The army and corps plans called for a rapid and deep attack, and the division received very little additional artillery support from the French Army.

In terms of liaison, the division never caught up after the disorganized initial attack. Colonel Malone first learned of the success of his regiment in the initial assault when German prisoners streamed past his command post.[22] The division did not have the materials it needed to maintain communication throughout the advance. For example, it was not provided with telephone wire, and the divisional horses, with the exception of those in the artillery brigade, were still lost in the French transportation network when the division attacked. This meant that runners were the only method of maintaining liaison between the front and commanders in the rear. Despite these communication problems, the division consistently accepted corps orders and continued to attack even though its forward units were not prepared to do so. The commander of the 3rd Brigade, General Hanson Ely, admitted that the evening attack on 18 July was "launched in a rather ragged manner." In fact, that is an apt description for every attack after the initial assault.[23]

The lack of preparation for subsequent attacks was nowhere more evident than in the poor coordination of artillery support, which was the second great weakness of the operation. After the initial assault, the leading infantry regiments were never given the opportunity to wait for the forward displacement of the divisional guns or sufficient time to coordinate an effective artillery fire plan. Colonel Upton reported that after his troops passed the limit of the rolling barrage, "there was absolutely no control of artillery fire by the troops in line or by the commanders." Here, too, the breakdown in liaison had a negative impact on the attacks. The light guns usually kept pace with the infantry regimental command posts, which meant that the artillery battalion commanders were as much in the dark about the location of the forward lines of the leading infantry elements as were the infantry regimental commanders. General Ely stated that "owing to the broken nature of the advancing lines and the lack of knowledge of the positions of [enemy] machine guns . . . and even at times of our own infantry, the assistance of the artillery was not as great as it otherwise no doubt would have been." The tendency was therefore to use the forward guns for counterbattery work, as that did not require precise information on the division's forward infantry positions.[24]

The division did a fair job of advancing what little artillery it had available. Batteries from the light gun regiments were in advanced positions in time to support the evening attack on 18 July, as well as on 19 July. Unfortunately, the division commander, General Harbord, prodded by the French Tenth Army and XX Corps, never gave the division the time it needed to catch its breath, reestablish liaison among its infantry units and artillery batteries, and coordinate an effective fire plan. In the end, the division was forced to fight in a mobile, "open warfare" style, and as it did so, it proved incapable of effectively employing its artillery forces in a manner that adequately supported the infantry.

One can summarize the three main attacks of the 2nd Division in the Aisne-Marne counteroffensive by analyzing three critical factors: the degree of surprise attained in the attack, the amount of time granted to prepare for the attack, and the quality of the artillery support given the infantry. In the initial assault, the division had the full effect of gaining near-total surprise on a tired, overextended, and poorly fortified enemy. This overshadowed the minimal amount of preparation time given to the division (i.e, to the division staff and artillery brigade) and the mediocre artillery support. These factors resulted in a great initial success, with moderate casualties. However, a reduction in the effectiveness of the attacks and an increase in the casualty rate were both apparent by the latter stages of the initial assault.

The second attack, on the evening of 18 July, was characterized by a greatly reduced level of surprise, virtually no time for the attacking units to prepare for the assault, and very weak to nonexistent artillery support. This attack resulted in some success, as the 2nd Division was the only division in the XX Corps to make any gains that evening, but the casualties were very heavy. Without a doubt, the missing machine-gun and 37mm gun units weakened the ability of the infantry to fight its way forward without sufficient artillery support.

By the third attack, on the morning of 19 July, the defending Germans had regained their composure, and reserves had been rushed to the area. The impact of the initial surprise was gone, and the Germans fully expected a continuation of the offensive. French army and corps commanders directed the division to coordinate the attack in less than two hours, most of which time was spent advancing the reserve infantry regiment to the attack positions. The artillery support plan called for one hour of preliminary fire from only a portion of the divisional guns, and, possibly due to the lack of time to perform all the required calculations, did not direct that a rolling barrage accompany the infantry. While the success of the attack was moderate at best, the casualties were excessive. On 19 July, after Lieutenant John D. Clark advanced his battery of 75s through the division sector, which was strewn with the masses of American dead, he wrote, "I am just beginning to realize what war really is."[25] In his learning Lieutenant Clark was in many respects joined by the rest of his division.

The division left Soissons, as it had Chateau-Thierry, battered but victorious. In both sets of operations, the 2nd Division fought through to some level of tactical success and helped make an impact of strategic importance. At Chateau-Thierry, the 2nd Division bolstered a worn-out and demoralized French Army, and by aggressively taking the offensive it proved to both the Allies and the Germans that the American Army, arriving in France by the thousands every day, was a force to be reckoned with on the battlefield.

At Soissons, the 2nd Division played a key role in a successful offensive that might have been the turning point of the war. The division's nine-kilometer advance on 18 July was the greatest single-day advance by an Allied division thus far in 1918. It led to the forced evacuation of the great German salient south of the Vesle River, drew German reserves away from the current German offensive near Reims, and helped destroy plans for a future offensive against the British Expeditionary Force (BEF) to the north. From 4:35 AM on 18 July until the end of the war, the German Army was forced to remain on the defensive, and it would move in only one direction—backward.

Nevertheless, the division achieved these victories at a great price. When it left Soissons, the 2nd Division, which had numbered 28,000 men in late May, had suffered a total of 13,719 casualties in killed, wounded, and missing or taken prisoner. Many factors beyond the control of the division contributed to those heavy losses, such as the hurried manner in which French commanders ordered the 2nd Division to attack, the lack of artillery augmentation from the French army and corps artillery forces, the lack of gas shells provided to the division near Chateau-Thierry, the near-total level of German air supremacy during every operation except the attack on Vaux, and the lagging advance of the French divisions on both flanks near Soissons, which forced the 2nd Division to hold an exposed salient position for nearly thirty-six hours. Some of these factors significantly decreased the division's combat effectiveness, and each led to increased casualties. Considering the total impact of all these negative influences, it is surprising that the 2nd Division was as successful as it was in its initial operations.

Despite these external factors, which decreased the division's effectiveness, it is also clear that the division did not consistently demonstrate the will or the skill to fight in a manner that maximized the use of all the division's available firepower. At Belleau Wood the division repeatedly attempted to make infantry attacks that were not adequately supported by artillery. However, these attacks, which were urged by General Harbord, reflected the approved AEF infantry-focused doctrine at the time. In fact, prior to joining the 4th Brigade in May, Harbord had spent his entire tour in France as Pershing's chief of staff, in which capacity he appears to have become a devotee of his commander-in-chief's concept of "self-reliant" infantry. Thus, it was Harbord's decision to return to the infantry-focused attacks, after the previous failures with that method and the successes of the artillery-centered approach, that most convincingly demonstrated that he was steeped in the doctrine and style of warfare that Pershing advocated.

The losses suffered in the three-week struggle to take Belleau Wood did appear to teach the 2nd Division a lesson, as it proved so well at Vaux. There the division demonstrated that while some of the officers might have been devoted to Pershing's infantry-focused style, others saw the benefits of the more conservative, artillery-centered approach. The division also demonstrated that it possessed the skills necessary to effectively plan and execute limited set-piece attacks that maximized the use of artillery.

At Soissons, the division was forced by the French Army to fight in the style Pershing advocated, a style that was based on surprise, mobility, and self-reliant infantry, and that afforded the infantry only occasional support from the artillery. In the fighting there, the 2nd Division confirmed that it did not possess the skills necessary to effectively fight in an "open warfare" environment. While the 2nd Division was able to successfully advance, it did so only because of the near-total level of surprise that was achieved, the weakness of the defensive positions opposing it, and the tremendous courage and spirit of the American infantrymen, who were willing to continue the attacks despite the terrible casualties that were occurring all around them.

If, after the operations near Chateau-Thierry, there was any doubt in the division as to the importance of overwhelming artillery support, the attack near Soissons went a long way towards expunging it. While Soissons had shown that a surprised defender, in weak defenses, could be initially overtaken even without great artillery support, it also proved that as the attack continued, and the effects of the surprise wore off, the successes would lessen and the casualties increase. Most of the officers and men who survived these early operations learned that the use of overwhelming artillery support was the only way to attack and succeed without excessive casualties.

Though the 2nd Division would henceforth make better attempts to maximize artillery support to cover its attacking infantry, in many respects it would be fighting an uphill battle. Pershing and his staff at American GHQ continued to look forward to an aggressive offensive in which his divisions would break into the open and fight in a mobile, infantry-focused style. After reviewing the July operations of all American divisions, Pershing's staff concluded, "The rifle again proved to be the chief weapon of the infantry soldier." Despite the lack of any effective artillery support in the later stages of the Soissons attack, the report claimed, "The regulation methods of handling artillery in open warfare were found to be sound and capable of execution."[26]

The operations of the 2nd Division near Chateau-Thierry and at Soissons had demonstrated that the American divisions did not possess the capability to effectively attack without using overwhelming artillery support to cover the infantry throughout the advance. When the 2nd Division did arrange adequate artillery support, as it did occasionally at Belleau Wood, initially at Soissons, and completely at Vaux, it demonstrated the capability to be surprisingly effective. The division showed that it possessed the skills necessary to carry out well-prepared, artillery-centered, set-piece attacks with great success. The problem with these conclusions was that they conflicted with the approved AEF doctrine, which stressed the need for the "self-reliant" American infantrymen to be aggressively employed in attacks that could achieve decisive results. While the 2nd Division might have learned its lessons, General Pershing and GHQ proved they had not yet learned theirs.

NOTES

1. Robert L. Bullard, *American Soldiers Also Fought* (New York: Longmans, Greene and Co., 1936), p. 13.

2. Journal of Operations, 2d Division, 15–17 July, *RSD*, vol. 6; Spaulding and Wright, *The Second Division*, pp. 105–8.

3. Spaulding and Wright, *The Second Division*, pp. 105–8; Harbord, *Leaves*, p. 318.

4. Spaulding and Wright, *The Second Division*, p. 108; Harbord, *American Army*, p. 317.

5. Spaulding and Wright, *The Second Division*, p. 108.

6. Harbord, *American Army*, p. 323–26; Spaulding and Wright, *The Second Division*, pp. 111–13.

7. ABMC, *American Armies*, pp. 31, 33.

8. Operations Memorandum No. 1, 2d FA Brigade, 17 July 1918, *RSD*, vol. 2.

9. Operations Memorandum No. 1, 2d FA Brigade, 17 July 1918, *RSD*, vol. 2.

10. Operations Order No. 227, XX Corps, 16 July, in 2nd Division Historical File, Entry 1241-32.7, RG 120, NA; Operations Memorandum No. 1, 2nd FA Brigade, 17 July 1918, *RSD*, vol. 2.

11. Report of Major General Charles P. Summerall to Deputy Chief of Staff, AEF, Subject: "Artillery," 5 January 1919, V Corps Historical File, Entry 1118, RG 120, NA.

12. An American officer serving with the French XX Corps reported that "traffic control, strictly speaking, does not exist . . . trains are tangled up in all directions, nobody issuing any orders on the subject . . . Faite Road has been completely blocked since 9 am yesterday, resulting in considerable delay in the arrival of the American 2d Division on the line." Report of Major Roy, Liaison Officer, XX Corps, 5 AM, 18 July in "French Tenth Army, G-3 Messages, July 18, 1918," Entry 318—30.1-30.9, RG 165, NA; "Report of Action of 3rd Brigade, June 1st to July 15th/18," 16 August 1918, *RSD*, vol. 7.

13. Field Message, 18 July 1918, 5:40 AM, Colonel Upton to Brigadier General Hanson Ely, Commander, 3rd Brigade, *RSD*, vol. 4.

14. Spaulding and Wright, *The Second Division*, pp. 117–19; "Colonel R. M. Johnson's comments on the Second Division at Soissons," in Folder 10, 2nd Division files, G-3 Reports, AEF/GHQ, Entry 270, RG 120, NA.

15. "Report of Operations, July 17–18–19, 1918," 23rd Infantry Regiment, 21 July, in Folder 70, 2nd Division files, G-3 Reports, AEF/GHQ, Entry 270, RG 120, NA; Spaulding and Wright, *The Second Division*, pp. 120–21.

16. "Report of Operations, July 17–18–19, 1918," 23rd Infantry Regiment, 21 July, in Folder 70, 2nd Division files, G-3 Reports, AEF/GHQ, Entry 270, RG 120, NA; "Report of Second Attack made NE of Vierzy on 18th–19th of July, 1918," 23 July 1918, 2d Battalion, 5th Marine Regiment, Folder 88, 2nd Division files, G-

3 Reports, AEF/GHQ, Entry 270, RG 120, NA; Spaulding and Wright, *The Second Division*, pp. 123–24.

17. "Report of Operations, July 17–18–19, 1918," 23rd Infantry Regiment, 21 July, in Folder 70, 2nd Division files, G-3 Reports, AEF/GHQ, Entry 270, RG 120, NA; Message, Colonel Upton to General Ely, 19 July, in *RSD*, vol. 4.

18. "A Brief History of the Sixth Regiment, USMC," typescript in Folder 90, 2nd Division files, G-3 Reports, AEF/GHQ, Entry 270, RG 120, NA; "Report," 2nd Division Commander to XX Corps Commander, 19 July 1918, in *RSD*, vol. 6.

19. Report of the 2nd Division Commander (Major General John A. Lejeune) to Asst. Chief of Staff, G-3, GHQ, AEF, Subject: "Operations Information," 30 December, 1918, in Reports and Other Records Regarding the History of AEF Units, Divisional Skeleton Histories, Entry 443, RG 120, NA.

20. "Report of Operations, July 17–21, 1918," in *RSD*, vol. 6.

21. "Report of C.O. Moroccan Division, of Operations July 8–22, 1918," in File 587–30.4, Entry 403, RG 165, NA.

22. "Report of Operations, July 17–18–19, 1918," 23rd Infantry Regiment, 21 July, in Folder 70, 2nd Division files, G-3 Reports, AEF/GHQ, Entry 270, RG 120, NA.

23. "Report of Operations, Third Brigade July 17th to 21st, in the attack of Twentieth Army Corps 'X' French Army," 27 July 1918, in *RSD*, vol. 7.

24. "Report of Operations, Third Brigade July 17th to 21st, in the attack of Twentieth Army Corps 'X' French Army," 27 July 1918, in *RSD*, vol. 7; "Report on Attack, 1930, 18 July," 9th Infantry, in Folder 59, 2nd Division files, G-3 Reports, AEF/GHQ, Entry 270, RG 120, NA.

25. Captain John D. Clark, Diary, 19 July 1918, 15th FA folder, 2nd Division boxes, World War I Manuscript Collection, USAMHI.

26. GHQ, AEF, "Notes on Recent Operations, No. 1" (France, 7 August 1918), in Report of G-5, Appendix 34, Folder 249, Commander-in-Chief Reports, GHQ, Entry 22, RG 120, NA.

4

St. Mihiel: The 2nd Division in the First American Army

By 11 a.m. September 12, the 2nd Division "had taken Thiaucourt, and very soon afterward it captured the high entrenched ground to the north of Thiaucourt. From that moment the Saint-Mihiel salient was no longer tenable by the enemy."[1]

Lieutenant General Hunter Liggett
Commanding General, American First Army

On 7 August, General Pershing mused to one of his staff officers, "It seems to me that perhaps we are losing too many men. . . ."[2] This was the AEF commander-in-chief's response to the American casualty figures for the Aisne-Marne counteroffensive, which had ended on the previous day. All told, the eight AEF divisions that participated in the reduction of the Aisne-Marne salient suffered some 50,000 casualties in the fighting between 18 July and 6 August.[3] And these casualties were suffered in fighting that was as close to Pershing's vision of "open warfare" as anything the Allies had seen in over three years on the Western Front. Pershing, therefore, directed his staff to investigate the methods of attack utilized in the recent operations and to make recommendations as to what tactics and techniques needed to be employed in order to make future American attacks less costly.[4]

While Pershing's staff at GHQ did their research, the 2nd Division continued its move into a quiet "rest sector" in the Nancy region (near the St. Mihiel salient), occupying the sector on 9 August and serving there under the French XXXII Corps for about two weeks. As expected, there was little hostile activity of any kind during the period, and the

Figure 4.1. Brigadier General John A. Lejeune, USMC. Later major general, Lejeune was commander of the 2nd Division during the St. Mihiel, Blanc Mont, and Meuse-Argonne attacks. Courtesy U.S. Army Military History Institute.

division concentrated on absorbing replacements and digesting the lessons of its recent operations.[5] After leaving the front line sector, the division was assigned to the U.S. I Corps and put through a week of training that culminated in a set of divisional maneuvers on the last two days of August. While not all divisional troops took part in the maneuvers, all unit commanders and their staffs down to the company level worked through a "terrain exercise in command, liaison and control of units."[6]

General Liggett, the I Corps commander, told General Lejeune, the new commander of the 2nd Division (see Figure 4.1), that "judging from the experience gained in the great offensive north of the Marne, it appears that the question of liaison is the subject that should receive very careful attention" in divisional training[7]. Therefore, in the maneuvers the division would be called on to advance a distance of ten kilometers, all the while affording commanders "practice in approaching strong points" and "in controlling men as they pass through woods." The division advanced according to a programmed timetable, similar to the attack near Soissons, where the units moved forward at a rate of 100 meters in three minutes all the way to an intermediate objective, and then at a slightly slower rate until completing the movement. During the advance, umpires occasionally handed out "special situations" to battalion and company commanders, and then, depending on the commander's actions, determined whether or not the unit would be permitted to continue forward. Along with representatives from all four infantry regiments and the machine-gun units, one battalion from each of the light field artillery regiments was included in the maneuver. To accentuate the "open warfare" nature of the maneuver, commanders were allowed to use "all means of communication . . . except the telephone." The attacks near Soissons had shown that the divisional infantry and artillery units needed to improve their ability to communicate with one another while advancing, and these maneuvers were a clear attempt to afford that opportunity.[8]

Shortly after the division completed the maneuvers, the commander of the 2nd Field Artillery Brigade, General Bowley (see Figure 4.2), released a new plan of liaison to improve communication and coordination between the infantry and artillery units of the division. During future operations, both the division headquarters and the artillery brigade headquarters would maintain "advance message centers," well forward of the respective headquarters. These sites would be connected by phone lines to each other and to both headquarters, adding security and flexibility to the communication system. One liaison officer and two couriers from each light artillery regiment would be kept at the artillery brigade's advance center, charged with maintaining liaison with the brigade representatives. In order to stay abreast of the infantry, liaison officers would also be assigned to each attacking infantry battalion, instructed to use the wireless maintained by the infantry to make "emergency contact" with their own regimental commanders.[9]

New ideas were taking shape throughout the AEF. American staff officers were busy trying to learn lessons not only from their own operations, but also from the operations of the other armies—especially that of Germany in the 1918 spring offensives. While the British and French viewed General Ludendorff's massive attacks as near disasters, Pershing took a certain amount of satisfaction in seeing that "open warfare," in one form or another, had actually been brought to the Western Front.[10] Pershing and his staff pored over the reports of the German attacks, hoping to glean the significant tactical details that had proven so capable

Figure 4.2. Brigadier General Albert J. Bowley. He was commander of the 2nd Field Artillery Brigade from the final taking of Belleau Wood to the end of the war. Courtesy U.S. Army Military History Institute.

of breaking open the front lines and restoring some ability to maneuver on the Western Front battlefield.

One of the tactical innovations that the AEF borrowed from the German Army was the parceling out of artillery detachments, either in individual pieces or single batteries, to the assaulting infantry battalions. This was one way of reducing the problem of maintaining liaison between the front-line infantry's troops and their supporting artillery. The use of accompanying guns reduced the infantry's reliance on a precarious telephone connection, or a slow runner, back to some regimental artillery command post, which would first need an accurate description of the hostile target and then often demand that the infantry pull away from the target to give the distant guns a safe margin for firing. With accompanying guns, on the other hand, the company and battalion commanders could call back just a few yards, point to the targets, and order for fire to be delivered exactly where it was needed. There was no long halt in the infantry advance—only as long as it took the gun to set up, aim, and fire. If the artillerymen were aggressive enough with their guns, the infantry would have the support they needed without the initiative-wrecking reliance on prearranged barrage lines. Furthermore, it appeared that this method of support would become more useful as the fighting became less trench-oriented and more in line with "open warfare." It seemed the perfect method of artillery support for Pershing's infantry.[11]

This new idea was quickly incorporated into AEF doctrine. Officers at the AEF General Staff College were taught that "artillery must accompany the infantry, not only by fire, but with the guns themselves in order to insure constant support, whatever may be the depth of the penetration." The use of accompanying artillery meshed perfectly with another concept that the AEF saw the Germans use effectively—infantry infiltration tactics. Pershing wanted no more neat waves of walking infantrymen in AEF attacks, which presented easy targets for enemy machine guns. AEF doctrine directed the infantry to advance between the enemy's strongest points, much like the German Army's stormtroop tactics. This would allow friendly artillery a free hand to concentrate on the strong points. However, any smaller nests of machine guns in the areas between the strong points would "have to be overcome by infantry alone, with the help of its own weapons (37mm) and of the accompanying artillery."[12] The primary difference between Pershing's infiltration tactics and those of the German stormtroopers was in the level of emphasis placed on using firepower to effect the penetration. While American infantry battalions were to be supported by machine guns, light mortars, and 37mm cannons, Pershing continued to insist until the end of the war that the rifle and bayonet were still the primary weapon of the infantryman. Conversely, the German stormtroopers based their attacks on the coordinated use of the most powerful infantry weapons available, primarily light machine guns, but also flamethrowers and grenade launchers.[13]

Pershing officially spelled out these tactical innovations on 5 September 1918, when AEF GHQ published his *Combat Instructions*. This pamphlet was the result of the research done by Pershing's staff following his concern about excessive casualties in the Aisne-Marne counteroffensive. The instructions began with a critique of the tactics used in the previous attacks, specifically, that as far as the infantry was concerned, there was "almost never any attempt to maneuver, that is, to throw supports and reserves to the flanks for envelopment." Pershing insisted that "the attempt to use trench warfare methods in an open warfare combat [sic] will be successful only at great cost." Instead, "where strong resistance is encountered, reinforcements must not be thrown in to make a frontal attack at this point, but must be pushed through gaps created by successful units to attack these strong points in the flank or rear."[14] These tactical concepts were fairly advanced, approximating the most progressive ideas being used in the German Army, as well as by the best British, Canadian and Australian divisions.[15]

To facilitate this new approach in "open warfare," the divisional artillery was to be divided into "two classes: (a) that retained by the division commander under the command of the artillery brigade commander, and (b) that assigned to infantry units under the command of the infantry commanders." Just how much artillery was placed in each class was a matter to be determined by the division commander in accordance with the situation. Despite his enthusiasm for this innovation, Pershing admitted that the new artillery dispositions each had strengths and limitations. "The retention of a large proportion under the division commander permits rapid and powerful concentration on decisive points anywhere on the division front. But unless communications are perfect and information complete, it renders the close support of local actions difficult." As for the use of accompanying guns, Pershing claimed that it:

> binds such artillery support closely to the infantry it is supporting, and gives the infantry commander a powerful combination of arms with which to handle local situations without the loss of time. On the other hand, it tends to lessen the power of artillery concentration of the division as a whole and may render the infantry unit clumsy and immobile. Moreover, it demands a high degree of decision and initiative on the part of both the infantry and artillery commanders immediately involved.[16]

As to how commanders should organize this artillery, Pershing suggested that "a suitable proportion" of accompanying artillery would be achieved by breaking up one battalion of 75mm guns (three batteries, four guns each) for each attacking infantry brigade. At that point the artillery was again divided into classes. Two of the batteries were to be maintained as such (called "infantry batteries") and used as whole batteries to "attack stronger points of resistance, defense against local counterattacks [sic]," and to fire on hostile reserves. Fire was to be accomplished by using direct observation from forward positions that were occupied "with rapidity and boldness." The third battery was to be broken up into individual guns and assigned to front-line infantry battalions for attacking machine guns, tanks, and strong points. Fire from these guns was to be "direct" at a range of from 500 to 1,500 meters. The guns were to be "moved horsed, unlimbered under cover, and run forward by hand." As for protection for these detachments, the crews working the guns were to utilize all possible concealment, make use of the natural features of the ground, rely on their mobility, and to always keep in mind the relatively "small target" that would be opposing them. In the end, when determining the actual displacement of the guns for battle, the artillery brigade commander was "responsible that the full power of his weapons is developed and aggressively used."[17]

These artillery employment concepts, as with the infantry infiltration ideas, approximated the cutting edge of attack doctrine on the Western Front in 1918. However, two significant problems promised to work against AEF units as they attempted to implement the doctrine on the battlefield. First, equipment weaknesses were sure to present problems. The AEF did not even possess a good light machine gun or automatic rifle for the infantry to use with its infiltration tactics. The old French chauchat automatic rifle simply did not measure up to the impressive British Lewis Gun or the German MG08/15 and MP18/I machine guns.[18] As for using accompanying artillery with the most forward infantry units, the 75mm gun, while light by World War I standards, still needed horses to pull it and its shells through the combat zone. Typically fire fights that were too hot for attacking infantry units to handle on their own were even more damaging on the men, horses, and equipment of artillery batteries that needed to set up out in the open. Second, successfully carrying out either the infantry infiltration or mobile artillery tactics required a great deal of highly specialized training, and the AEF did not have the time or instructors needed to run effective training programs for the entire AEF in each of these areas.

While these new ideas were being disseminated, Pershing and his staff were finishing preparations for the first real American offensive of the war, the attack on the St. Mihiel salient. While AEF units had played a small part in stopping some of the German Army's spring offensives, and a more significant role in the Aisne-Marne counteroffensive, the reduction of the St. Mihiel salient was to be an AEF-dominated operation. This was going to be the first attack made by the newly created American First Army, and it would entail the use of three American and one French army corps. Although American corps had existed in an administrative capacity for months, and had even commanded some troops in battle in July and August, they never possessed all of the units required to effectively undertake active operations at the front. In order for the St. Mihiel attack to be successful, the American First Army had to develop the huge corps and army artillery organizations required to gain superiority over the German guns in the sector.

However, the AEF was clearly deficient in the heavy, long-range guns needed to create the corps and army artillery organizations, as well as in the trained men who were needed to operate and command them. The answer was to turn again to the French Army. Colonel Conrad Lanza, who served as the chief of operations for the First Army Artillery and was later given the task of writing a history of the organization, stated that "at times the French army furnished more than half of the total artillery in the First Army." Not only did the French Army provide the necessary guns and ammunition, it also furnished balloon and airplane units for use in gaining air superiority, and in conducting the important reconnaissance, intelligence-gathering, artillery-spotting missions. Finally, it provided several of the corps chiefs of artillery along with their associated staffs. In short, though the artillery of the First Army was ultimately an American organization under American leadership, Colonel Lanza could assert that "without the help of the French Army, the American Army could not have functioned as far as artillery is concerned."[19]

Despite this pervasive French influence, it was up to the senior American officers to develop a comprehensive artillery system for the First Army, ranging from the divisional artillery brigades to the massive corps and army artillery groups. In organizing the corps and army artillery organizations, the overriding goal was to develop a system that would effectively dominate the enemy artillery in the area, which meant that the system had to be effective in counterbattery operations. After months of studying and working with the French and British counterbattery systems, throughout August and the first weeks of September the First Army built a system that Colonel Lanza described as "a compromise" between the two, in which the best from each system was incorporated.

Pershing assigned all counterbattery operations directly to the corps artillery forces. One officer on each corps artillery staff was designated as the counterbattery officer, and was given his own observation posts, flash and sound ranging sections, as well as a number of designated aircraft to observe and photograph enemy artillery positions. While the corps was authorized to order batteries from its divisional artillery brigades to contribute to the counterbattery effort and could also request special assistance on specific targets from army artillery groups, in the end it was a corps responsibility to locate and neutralize enemy artillery.[20]

That the Americans were seeking to just neutralize enemy batteries, and were not intent on destroying them, is an important point. The concept of neutralization, instead of relying on destruction, was another lesson AEF officers had picked up from their study of the tactically successful German attacks in the spring of 1918. Until the massive German attack on 21 March 1918, counterbattery operations on the Western Front preparatory to a major offensive were lengthy affairs, in which the attacking force would spend days and even weeks methodically attempting to locate and destroy every enemy gun in the attack sector. During this plodding artillery preparation, the enemy was not only alerted to the upcoming attack, but had a number of days in which to prepare its defensive plan, move new troops into reserve positions, and bring in new guns that would remain hidden until the infantry attack started. While this system succeeded in destroying some targets, it was never able to destroy enough of them before the start of the offensive to prevent the attacking infantry from suffering heavy losses from the enemy artillery fire. During its successful attacks in the spring, the German Army used an artillery plan that was based on the premise that enemy artillery needed only to be neutralized during the attack, and not physically destroyed. Guided by this belief, the German Army was able to reduce the preliminary artillery preparation by well over 95 percent, from a matter of days to a period of about four hours.[21]

This was the guiding tactical philosophy behind the buildup of the corps and army artillery forces of the First Army prior to the St. Mihiel attack. The general plan for an offensive near the St. Mihiel salient had existed from the summer of 1917, when Pershing and his staff members looked at maps of the Western Front in search of a sector that the AEF could make its own. The salient itself was a thirty-kilometer deep bulge that extended into the French lines approximately twenty kilometers south of Verdun, which had been under continuous German control since they seized it in September 1914. Numerous French attacks against the massive protrusion throughout 1915 had all failed. After the Germans used the position to launch attacks during its assault on Verdun in 1916, the salient quieted down into a "rest sector."[22]

When Marshal Foch, the Allied generalissimo, met with Haig, Pétain, and Pershing in late July to discuss strategy, the reduction of the St. Mihiel salient was one of four limited counteroffensives on his agenda. By the end of August, after successes in the Aisne-Marne and Amiens offensives and the subsequent German withdrawal from the Ypres-Lys salient, the St. Mihiel operation was the last of the four remaining.[23] By then, Pershing's staff had been working on plans for the St. Mihiel attack for weeks, and had even completed a draft of the attack plan. While the main objective of the attack was the reduction of the salient, Pershing had initially planned to follow up this limited operation with a full-fledged advance toward the fortress city of Metz, some twenty kilometers beyond the base of the salient. Pershing intended for this attack to be the beginning of the decisive American offensive that he hoped would bring about the end of the war.[24]

However, on 30 August Foch informed Pershing that he wanted the St. Mihiel attack to be no more than a limited operation, so that the AEF could play a major role in a general offensive to take place in late September that would, with the coordinated advances of the French and British armies, drive back the giant bulge of German forces that extended into northeast France and Belgium. While at first dismayed by the suggestion, Pershing acquiesced after being assured that he would be able to command an independent American army, composed of American forces, in the upcoming offensive.[25] Thus, Pershing's first opportunity to push his forces into an unlimited offensive was delayed. The American Army's first great attack would be a giant set-piece operation.

Plans for the reduction of the salient were rapidly finalized, and divisions were designated and prepared for the attack. The defining feature of the plan was its insistence on absolutely overwhelming force. The plan called for fourteen American and French divisions, in four corps, to attack from both sides of the salient. This meant that some 660,000 Allied troops (550,000 American and 110,000 French) were to participate in an attack on an opposing German force within the salient that was estimated by American intelligence experts to number just 23,000.[26] The Allied infantry was to be supported by nearly 1,400 French, British, and American aircraft under the command of American Colonel William "Billy" Mitchell. While Pershing did not receive any of the 150 heavy tanks he requested, he did obtain 267 of the 300 light and medium tanks for which he asked.[27]

The artillery force supporting the attack was also massive, guaranteeing an overwhelming superiority of fire for the initial assault: 3,010 artillery pieces of all calibers and over three million rounds of ammunition were to support the assaulting troops.[28] The initial artillery employment plan called for a twenty-two-hour artillery preparation. However, in the interest of maintaining surprise, Pershing went so far as canceling the preliminary bombardment on some intermediate plans. Eventually, after reviewing the formidable defenses protecting the salient and receiving word that no heavy tanks were available, Pershing agreed to an intense four-hour preliminary bombardment.[29] This would give the supporting artillery a jump on neutralizing the enemy guns and destroying enemy defenses while not giving the enemy sufficient time to make any substantial alterations to its plan of defense.

The 2nd Division was ordered to advance on the western edge of I Corps, which was to attack at the southeastern base of the salient (see Map 4.1). The division was to occupy a front of just two kilometers, with the 5th Division on its right (east) and the 89th Division of the US IV Corps on its left (west). The I Corps attack plan directed its divisions to make three strong attacks over a two-day period. The 2nd Division's attack was to begin with the infantry, supported by forty-five light and eighteen medium tanks, following a rolling barrage that advanced 100 meters every four minutes. This would take them through the first five kilometers to the "first phase objective," where the barrage would halt for a period to allow the infantry battalions time to regroup, and then advance again to the "first day's objective." Throughout this advance, the front of the 2nd Division was to expand to a width of four kilometers. The plan called for organized attacks to stop for the night after a one-

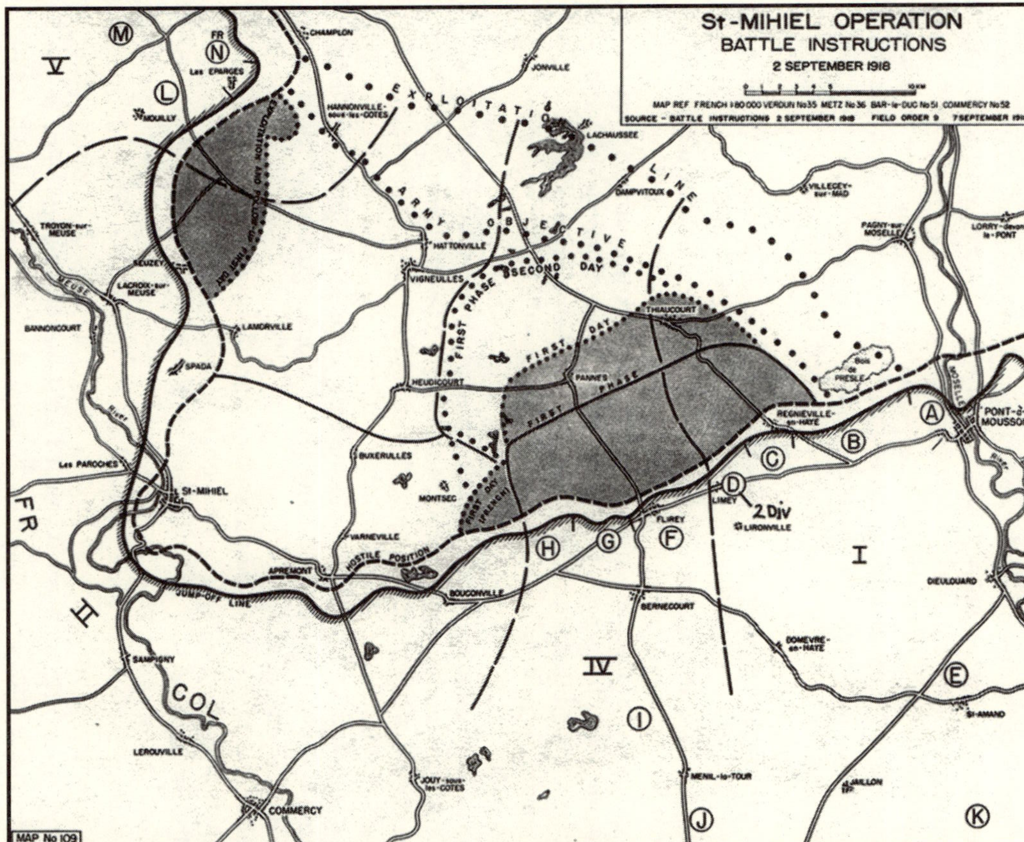

Map 4.1. Plan of attack for the First American Army in the St. Mihiel Offensive. The 2nd Division attacked on the western edge of the I Corps, located on the eastern portion of the St. Mihiel salient. Taken from *United States Army in the World War*, vol. 8, p. 179.

day advance of more than eight kilometers, a rather aggressive plan considering the strong defensive fortifications in the salient. On the second day, the division was to complete the drive with a short push to the final "army objective," and then concentrate on strengthening its position across from the new enemy line of resistance.[30]

While the plan attempted to script the whole attack in set-piece fashion, it promoted some of the "open warfare" concepts that Pershing described in his *Combat Instructions*. Assaulting infantry were told that "penetration will be sought by utilizing lanes of least resistance in order to cause the fall of strong points by out-flanking." Although the infantry advance was to be regulated by the rolling barrage, attacking units were directed that any "success" in the attack was to be "vigorously exploited." Just how this was to be done was left to the imagination of the division commanders and their subordinates.[31]

To accomplish its portion of the attack, the 2nd Division was augmented with two French regiments of 75mm guns and one French regiment of 155mm howitzers, bringing the total divisional artillery complement to 120 75mm guns and 60 155mm howitzers. All divisional guns were moved into position under the cover of darkness during the three nights preceding the attack. However, in the interest of surprise, all batteries were prohibited from registering their guns before the attack. They were forced to rely on the accuracy of their skills in "map firing," as well as in the great mass of fire that was put down, to provide sufficient support to the infantry. The prohibition of any preliminary artillery adjustment was another tactic borrowed from the German Army, as it had been a crucial element in achieving the level of surprise with which they attacked in the spring offensives.[32] From the corps and army artillery groups, the 2nd Division was allocated the dedicated support of three batteries of 105mm guns, four batteries of 120mm guns, and three batteries of 155mm guns.[33]

After a short, violent preliminary bombardment with high-explosive, shrapnel, and gas shells, the divisional artillery force was to carry the infantry through each of its objectives via the rolling barrage. While the corps artillery groups were to focus primarily on counterbattery operations during the attack, some were also designated to provide support to the infantry. These guns would fire concentrations in advance of the infantry "up to the moment when it might become a hindrance to the progress of the attack," which was designated as thirty minutes prior to the time the infantry was scheduled to reach that line. At that point the corps artillery was to lift and fire on more distant "designated objectives," all the while "being regulated by the march of the infantry." The corps plan also directed that division commanders attach one battalion of light guns to each attacking infantry brigade to be used as forward guns, presumably in the fashion outlined in Pershing's *Combat Instructions*.[34]

In compliance with the corps plan, General Bowley directed one battalion from each of his 75mm regiments to be "hitched and ready to move behind the infantry attack and occupy advanced positions." To facilitate this forward movement, each of these battalions was assigned its own designated caisson company from the divisional ammunition train, as well as a detachment from the divisional engineers. All other artillery battalions would be moved forward as directed by the artillery brigade commander. The rolling barrage was to be fired by the two remaining battalions from the 12th and 15th artillery regiments, along with one French regiment of 75s (the 247th). These units were ordered to ensure that their fire overlapped each other, so that there would be no gaps in the barrage. The other French regiment of 75s (the 49th) was directed to fire smoke screens and to join in on the standing barrage at the first phase objective. The two regiments of 155mm howitzers were to focus on cutting enemy wire (which V Corps orders designated as "the greatest obstacle to rapid progression"), as well as other "special targets."[35]

While Lejeune and Bowley did comply with orders to designate certain guns to move forward with the infantry, they apparently did not arrange to employ them in the manner described in Pershing's *Combat Instructions*. In fact, they only planned on employing those light battalions as accompanying guns for the very beginning of the initial infantry assault. During those first few kilometers, the advance battalions were ordered to respond to any special requests from the leading infantry units, with the battalion from the 15th FA assigned to support the 9th Infantry and the battalion from the 12th FA dedicated to the 23rd Infantry.[36] However, in order to extend and strengthen the rolling barrage, which was supposed to accompany the infantry for more than eight kilometers, Bowley had also directed those advance artillery battalions to eventually move forward to designated locations where they were to pick up the rolling barrage and fire it from the "first phase objective" to the "first day's objective." This was required because the portion of the rolling barrage beyond the "first phase objective" was out of range for many of the 75mm batteries that were to fire the barrage from the beginning of the attack. As long as the advance battalions took up positions far enough forward, they would be able to fire the barrage all the way to its completion, and even take part in the thirty-minute standing barrage at the end of the day's attack.[37]

By the early morning of 12 September, all was in place. In fact, in comparison to previous engagements, troop commanders in the 2nd Division felt extremely well prepared. General Ely, commander of the 3rd

Brigade, which was leading the assault, said that he had spent three days poring over maps, reviewing the actual terrain, and preparing himself and his unit for attack.[38] All troops were in their attack positions three hours prior to "H" hour. The guns designated to participate in the preparation and rolling barrage had been in or near their firing positions for days and had their limbers and caissons at their positions so that they could advance as soon as possible. The 155mm batteries of the 17th FA were positioned far enough forward (i.e., one to two kilometers from the front trenches) that they could fire all their missions for the first day without having to move forward. The two "accompanying" battalions were hitched and ready to move forward on the heels of the attacking infantry.[39] Artillerymen from the 2nd Trench Mortar Battery, whose nearly obsolete 58mm weapons were not included in the fire plan, were formed into two squads and readied to follow the supporting infantry units on foot in order to take over and put to use any serviceable enemy guns that were captured during the attack.[40] The entire plan of artillery support was briefed to all infantry brigade, regimental, and battalion commanders, who were instructed to call directly to any artillery battalion or regimental commander for extra fire support. Artillery liaison officers were detailed to the leading infantry battalions. All was ready, and General Lejeune voiced his confidence by asserting that his division was in "splendid condition."[41]

Promptly at 1:00 AM, 12 September, the artillery preparation began, and continued for the prescribed four hours. During this time, the corps and army artillery groups hit enemy gun emplacements, crossroads, command posts, and railroads, while the divisional artillery batteries focused on barbed wire entanglements, enemy strong points, and troop cantonments.[42] At "H" hour, 5:00 AM, the divisional 75s began the rolling barrage, which was preceded by smoke screens and concentrations of heavy fire from the 155s. Additional smoke was poured on enemy observation posts and machine guns. The smoke shells combined with heavy fog and mist to produce a blanket of cover for the attacking infantrymen.[43]

All initial indications pointed to an attack going almost perfectly. The leading infantry battalions found "very little resistance . . . in the German front trench system." While some machine-guns nests in the woods offered "stubborn resistance," all were flanked by small detachments of infantry, allowing the bulk of the forces to stay close to the rolling barrage and continue the advance. The first "intermediate objective," some two and a half kilometers past the jump-off, was reached at 7:00 AM and was "passed without a stop." Shortly after this time, the accompanying artillery battalions headed into "No Man's Land" and, with the help of the detached engineers, advanced toward their forward firing positions. Neither battalion was called on to offer any direct support to the attacking infantry units. By 10:00 AM, both leading infantry regiments, the 9th and 23rd Infantry Regiments of the 3rd Brigade, had hit the "first phase objective" just south of the town of Thiaucourt, reorganized themselves for the advance in the second phase, and were waiting anxiously for the standing barrage to begin rolling again at 11:00 AM. The leading artillery battalions were in advanced positions by 10:30 AM, and they promptly joined in when the rolling barrage began to advance again on schedule thirty minutes later. The first phase of the attack could not have gone much better.[44]

As the barrage rolled forward, the infantry continued its steady advance, taking Thiaucourt in time to capture a number of German artillery pieces loaded on a train just moments before it was sent into the German rear. Despite outstripping the 5th and 89th Divisions on its flanks, the 2nd Division met little resistance and continued its advance all the way to the "army (or second day) objective," which was reached before 2:00 PM on the first day. As the leading infantry units consolidated their new positions, they threw out "strong combat patrols" to the front to establish an outpost line. Having accomplished all that the I Corps had ordered, and done so well ahead of schedule, the division could claim that the operation was practically over and that it was a striking success. The division had advanced more than eight kilometers over a strongly fortified enemy position and captured over ninety artillery pieces as well as approximately three thousand prisoners. And all this was accomplished at a cost of casualties that were in the hundreds.[45]

As would be expected in an attack that was this successful, infantry reports and assessments of the artillery support were glowing. As for the preliminary bombardment, General Ely thought it was "very effective" because "enemy PCs [Post of Command] and many machine-gun nests and trenches were knocked out and the enemy's artillery fire was feeble." The commanders of the two assaulting infantry regiments agreed that artillery support of the infantry throughout the whole of the assault phase was "excellent" and that it was "all that could be asked for."[46] The rapidity of the advance, the numbers of enemy troops and material captured, and the slight casualties sustained in the initial assault, all confirm the infantry assessments.

However, General Bowley's report on the initial operations of his artillery brigade did not fail to critique the corps orders that had directed him to assign an artillery battalion directly to the attacking infantry units. While claiming that he had placed artillery units "under the orders of the commanding general of the leading [infantry] brigade," he asserted that "the co-operation between the Field Artillery and the Infantry of this division has been such that there has been no necessity for assigning Field Artillery units to the infantry. The 75m/m regiments have always been zealous in pushing forward in direct support of the Infantry." He further claimed that with his plan of artillery support, which called for the rolling barrage to cover the entire advance of the infantry, "the leading Infantry Brigade had the immediate support of two regiments of 75s, instead of one battalion as ordered." This comment reflected Bowley's belief in the importance of covering as much of the infantry advance with a mass of coordinated firepower as possible.[47] He did not care for the idea of dispersing his guns in a "one here, one there" fashion, which made it almost impossible to gather together the kind of massed firepower that had proven to be the only method of adequately supporting the infantry and reducing their casualties. Not only was he convinced that his method was more effective in controlling and concentrating the firepower of his guns, but he was surely influenced by the unimpressive infantry reports on the use of his "accompanying artillery." Colonel Edward R. Stone, the commander of the 23rd Infantry Regiment, reported that the "advance was too rapid for the accompanying guns to keep up with the attacking battalions." The commander of the 4th Marine Brigade, General Neville, reported, "with well trained artillery which responds quickly and accurately to the call of the infantry, accompanying guns are not necessary if special infantry weapons are properly used."[48] With such reports coming from the infantry commanders, combined with the apparent success of his lengthy rolling barrage, it is not surprising that Bowley resisted implementing "accompanying" artillery in the future.

While the initial advance had been a clear success, it did not signal the end of the engagement, and as with most attacks against the German Army in the war, strong counter-attacks occurred that afternoon. Unfortunately for the infantry of the 2nd Division, the artillery support needed to help repel these counterattacks was not as perfectly scripted as was the rolling barrage that had carried them that far into enemy territory. At 3:30 PM the 9th Infantry Regiment was hit hard by a German counterattack, which was supported by numerous machine guns and at least one artillery battery. For three hours the front-line infantrymen fought off these attacks without any artillery support since they had lost communication with their supporting artillery. Finally, at 6:30 PM one battery of the 15th FA "came into action" and assisted in suppressing the counterattack, but by then the 9th Infantry Regiment had "suffered more losses than during all the rest of the operation." A similar situation occurred on the left of the division front, where the 23rd Infantry Regiment reported that between 2:00 PM and 7:00 PM "liaison between the regiment and supporting artillery . . . did not exist." Colonel Stone reported that during this period it was "impossible" to get the artillery support required to hold the advanced positions.[49]

General Ely claimed that during that first afternoon the unanswered German artillery and machine-gun fire nearly forced his troops to retire to a reverse slope position to reduce casualties. The silence of the 2nd Division's artillery was all the more frustrating to Ely because he was convinced that some batteries were in positions far enough forward to have offered assistance. As far as Ely was concerned, the problem was primarily one of communication. He reported that the artillery liaison officers with his regiments "did not know where the guns were, and had no horses, telephones or other means of communicating with them." He believed that the problems were caused by batteries advancing into forward positions and then neglecting to report their positions to anyone. As he saw it, it was a liaison problem internal to the artillery brigade.

However, there really were two problems contributing to the lack of artillery support on the afternoon of 12 September. First, most of the batteries experienced "great difficulties" in getting through the numerous trench systems in the attack sector. Those that did get their guns forward had to struggle to move their ammunition across the vast wasteland of the previous "No Man's Land," an equally difficult task.[50] The slow progress of the artillery advance was aggravated by the faster-than-expected advance of the infantry in the latter stages of the attack. Second, those batteries that were able to set up within range of the forward infantry positions had trouble establishing their communications network. Not only did they have to establish and maintain contact with their regimental command post in the rear, they also had to find out the correct positions of the most forward infantry troops. Lejeune reported that,

> the apparent failure of a lack of artillery support during the afternoon of September 12th was due to the rapid advance of the infantry, and to the failure of obtaining prompt telephone connections. Artillery was in position to give prompt support but hesitated in doing so due to lack of exact knowledge of the infantry front line positions.[51]

Stated this way, the breakdown in liaison looks less like a problem internal to the artillery brigade, and more like one of communication between the infantry and the artillery. Nevertheless, the fact remains that at the end of the first day's advance, artillery support was needed, and it was not provided. The lack of it put the afternoon's gains in jeopardy and led to increased loss of life among the infantrymen.

The following day, the 4th Marine Brigade relieved the 3rd Brigade, and little else of consequence occurred. However, on 14 September the division complied with corps orders to push its front line closer to the enemy main line of resistance, which was now behind the Michel Stellung of the Hindenburg Line. These localized attacks, referred to as "strong patrols" in the operations reports, seem to have been undertaken without much artillery support. While the advance of the 5th Marine Regiment on the right went smoothly, the 6th Marine Regiment ran into serious opposition on the left. Though the forward positions were eventually taken, the 6th Regiment suffered approximately 500 casualties in the hastily developed attack.[52] The division was subsequently relieved on the night of 15 September by the 78th Division, ending the operation.

At face value, the St. Mihiel attack was a great success for the American First Army. Within two days it had completely eliminated the entire salient, liberating some 500 square kilometers of French soil that had been under enemy control since the first months of the war and freeing up a critical railway that connected some eastern industrial areas to Paris. Nearly 16,000 German soldiers were captured, along with 450 artillery pieces. All this was accomplished while sustaining a total of just 7,000 casualties in the assault phase of operation.[53] The 2nd Division had done its share. It had advanced further, and faster, than the divisions on its flanks, taken more than 3,300 prisoners, and captured more than 120 enemy guns. Its total casualties from 12 to 16 September totaled 1,551 soldiers. In these terms, it was a successful operation for the division as well.[54]

However, certain factors, if taken into account, would have tempered the enthusiasm with which the AEF, and the 2nd Division, viewed operations at St. Mihiel. Significantly, the German Army had taken the first steps towards a complete tactical withdrawal of the salient before the attack began. In spite of every attempt by the Allies to maintain operational security concerning the planned offensive, German forces suspected an attack on the salient. After briefly considering and then ruling out the option of making a spoiling attack, on 10 September Ludendorff ordered all German forces in the salient to execute a preprogrammed withdrawal to behind the well-fortified Hindenburg Line position at the base of the salient, called the Michel Stellung.[55] While no troops were removed in the first stages of the withdrawal, some heavy material was removed, and other changes were made to the defensive plan. Certainly, the German defenses in the salient were in some heightened state of flux when the attack came, and this hampered their defensive effort.[56]

This squares with the assessment of enemy resistance offered in the operations reports of the 2nd Division, I Corps, and just about the whole attacking force of the First Army. Nearly all reports confirmed that the enemy artillery response was surprisingly weak and that many of the strong points were evacuated in the face of the attacking troops. General Ely admitted that if certain machine-gun nests had been held "more valiantly," his infantrymen "would have had a hard time."[57] The relatively poor defensive showing by the German Army is an important consideration when evaluating the performance of the 2nd Division, and of the First Army, at St. Mihiel. The issue is not that the AEF might not have been able to eliminate the salient; a review of the comparative Allied and German forces available for the engagement compel one to agree with the German High Command that a defense of the salient was probably futile. The important point is that even with the disorganized German defense, there were indications that a breakdown was occurring in the 2nd Division, especially concerning its ability to continue the advance.

The difficulty of moving the artillery and supplies forward, the problems of locating and communicating the most advanced infantry positions, and the lack of time to thoroughly prepare coordinated fire support plans for the later operations, all pointed to a divisional attack that was losing, or that had already lost, much of the powerful momentum with which it had started the offensive. With this in mind, it appears that the decision to make the St. Mihiel Offensive a limited, set-piece attack was a beneficial one, even if made for different reasons. Considering the strength of the Michel Stellung, and the disposition and logistical condition of the 2nd Division, any continued advance would have required a buildup and preparation similar to that effected for the initial attack on 12 September, or would have forced the division to accept increasingly heavy casualties, with decreasing gains to show for them. The 2nd Division, and the American First Army, had hit the limit of its effectiveness. It accomplished the planning and execution for the set-piece attack admirably, but any attempt to continue the attack immediately would almost certainly have met with failure. General Summerall, the commander of the 1st Division at St. Mihiel, voiced a similar belief when he said that,

> On the third day, the resistance stiffened and it was unmistakable that the enemy had occupied his new line in considerable strength. . . . Even if the [American First] Army had not devoted the troops to other missions [i.e. to the Meuse-Argonne area], there would soon have come a period of temporary stabilization during which the enemy would have constantly strengthened his lines and we should have made preparations for an attack on his organizations in much the same manner as at the beginning.[58]

Not surprisingly, Pershing took a more optimistic view of the situation. After receiving numerous messages of congratulations from Allied and American political and military leaders, he stated that this attack had vindicated all he had said about the fighting value of the aggressive and individualistic American soldier. Musing to an aide just after the attack, Pershing said that America had "developed a type of manhood superior in initiative to that existing abroad, which given approximately equal training and discipline, developed a superior soldier to that existing abroad."[59] As for the attack itself, he stated that "without a doubt, an immediate continuation of the advance would have carried us well beyond the Hindenburg Line and possibly into Metz."[60] It is obvious that Pershing did not seriously consider either the "unexpectedly light" level of resistance offered by the Germans or the difficulties that his divisions were experiencing in the later stages of the operation. He was filled with confidence in his own leadership, his doctrine, and his army, and was eager to put it all to the test in the upcoming offensive in the Meuse-Argonne sector on 26 September. He believed that he would finally be able to demonstrate the superiority of his "open warfare" doctrine in an operation with comparatively unlimited objectives.

Despite the lower-than-average level of resistance of the part of the defending Germans, the operation was by all accounts a major victory. The St. Mihiel Offensive proved that the 2nd Division, and other AEF divisions, possessed the ability to successfully plan and execute limited, set-piece attacks, even on a large scale. However, the operation also exposed some of the AEF's weaknesses, most notably the problems associated with immediately following one great advance with further attacks.

NOTES

1. Hunter Liggett, *Commanding an American Army* (New York: Houghton Mifflin, 1925), p. 65.

2. Pershing, in a memorandum to his chief of staff, Major General James McAndrew, as quoted in James W. Rainey, "Ambivalent Warfare: The Tactical Doctrine of the AEF in World War I," *Parameters*, Journal of the U.S. Army War College (September, 1983), p. 41.

3. The eight divisions were 1st, 2nd, 3rd, 4th, 26th, 28th, 32nd, and 42nd Divisions. Counting American air and corps artillery units, Pershing stated that over 300,000 Americans took part in the offensive. See Pershing, *My Experiences*, II: 211.

4. Pershing memo, 7 August 1918, in Rainey, "Ambivalent Warfare," pp. 41–42.

5. Report of the 2nd Division Commander (Major General John A. Lejeune) to Assistant Chief of Staff, G-3, GHQ, AEF, Subject: "Operations Information," 30 December 1918, in Reports and Other Records Regarding the History of AEF Units, Divisional Skeleton Histories, Entry 443, RG 120, NA.

6. While the official records give no clear reason as to why most of the division's troops would not take part in the maneuvers, a few memoirs of young soldiers give hints. One young Marine implies that the division was too preoccupied processing, indoctrinating, and drilling the thousands of new replacement troops. See Elton E. Mackin, *Suddenly We Didn't Want to Die: Memoirs of a World War I Marine* (Novato, CA: Presidio Press, 1993), p. 133; Memo from Commanding General, I Corps to Commanding General, 2nd Division, 24 August, 1918, Subject: "Training." 2nd Division Historical File, Entry No. 1241—50.4-56.3, RG 120, NA.

7. Major General Lejeune, of the U.S. Marine Corps, replaced Harbord on 28 July as commander of the division. Harbord was hand-picked by Pershing to take command of the AEF Services of Supply; Memo from Commanding General,

I Corps, to Commanding General, 2nd Division, 24 August, 1918, Subject: "Training." 2nd Division Historical File, Entry No. 1241—50.4-56.3, RG 120, NA.

8. General Ely, the commander of the 3rd Brigade, twice referred to these maneuvers as a "rehearsal" for the upcoming attack at St. Mihiel. See "Report of Operations, Third Brigade, September 17th, 1918," *RSD*, vol. 7; Memo, 2nd Division, 29 August, 1918, 2nd Division Historical File, Entry 1241—50.4-56.3, RG 120, NA; Lieutenant William D. Bickham and Major William Burr, "Second Field Artillery Brigade History," undated typescript, 2nd FA Brigade Historical File, Entry 1246-11.4, RG 120, NA.

9. Memorandum, HQ 2d FA Brigade, 2nd Division Historical File, Entry No. 1241—50.4-56.3, RG 120, NA.

10. Smythe, *Pershing*, pp. 97–98.

11. For an explanation of the German Army's use of accompanying guns, see Zabecki, *Steel Wind*, pp. 43–44; for incorporation into American doctrine see Pershing, *Combat Instructions*, 5 September, 1918, G-5, GHQ, AEF Document No. 1348, pp. 1, 6.

12. Lecture at Army General Staff College, Conference 57, 3rd Course, "Artillery in the War of Movement," 1 August 1918, by Col. Koechlin-Schwartz. G-5 Schools File, Entry No. 362, RG 120, NA.

13. Bruce I. Gudmundsson, *Stormtroop Tactics*, p. 142.

14. Pershing, *Combat Instructions*, 5 September, 1918, G-5, GHQ, AEF Document No. 1348, pp. 1, 6.

15. For a detailed description of the most advanced tactical concepts in the Canadian divisions, see Shane B. Schreiber, *Shock Army of the British Empire* (Westport, CT: Praeger, 1997).

16. Pershing, *Combat Instructions*, p. 6.

17. Pershing, *Combat Instructions*, pp. 6–8.

18. The French Chauchat could only fire 250 rounds per minute, while the BEF's Lewis Gun could shoot between 500–600 rpm, the German MG08/15 between 450–550 rpm, and the MP18/I about 400 rpm. Furthermore, the often inaccurate Chauchat was notorious for jamming in combat due to muddy conditions or overheating, which occurred after firing as few as 300 rounds in a clean gun. For a comparison of World War I machine guns, see Robert Bruce, *Machine Guns of World War I* (London: Windrow and Greene, 1997). For a slightly more favorable perspective on the Chauchat, see Gerard Demaison and Yves Buffetaut, *Honour Bound: The Chauchat Machine Rifle* (Cobourg, ONT: Collector Grade Publications, 1995).

19. Conrad H. Lanza, *The Army Artillery, First Army*, p. ii. An undated typescript of this unit history, which was commissioned by the U.S. War Department, is held at the U.S. Army Military History Institute, Carlisle Barracks, PA.

20. Conrad H. Lanza, "Counterbattery in the A.E.F.," *FAJ* (September, 1936), pp. 455–56.

21. Added advantages of neutralization fire were that it required less accuracy than fire for destruction, could be fired at night and in bad weather, reduced the total number of shells that needed to be fired in a preparation, and therefore did not chew up the ground as much, making it easier for the infantry and artillery to move forward during the attack. Lanza, "Counterbattery in the A.E.F.," pp. 457–58.

22. Braim, *The Test of Battle*, p. 77. For a good description of the strategic factors involved with the reduction of the salient, see Trask, *Coalition Warmaking*, pp. 19–22.

23. Trask, *Coalition Warmaking*, pp. 94, 99.

24. Pershing, *My Experiences*, II: 252, 263.

25. Pershing, *My Experiences*, II: 243; Trask, *Coalition Warmaking*, pp. 103–104.

26. Pershing stated that the salient was held "usually by seven enemy divisions in the front line, with two in reserve." Trask asserts that the "effective strength" of the seven "on-line" divisions was 23,000 troops. The AEF estimated that the German forces could be reinforced by two divisions within two days, two more in three days, and by as many divisions as the Germans could make available in four days. *My Experiences*, II: 264; Trask, *Coalition Warmaking*, p. 106.

27. Pershing, *My Experiences*, II: 160–61. Braim, *Test of Battle*, p. 77.

28. Lanza, *The Army Artillery, First Army*, pp. 66–67.

29. Lanza, *The Army Artillery, First Army*, pp. 65. Pershing, *My Experiences*, II: 265.

30. Field Order No. 49, HQ I Corps, AEF, 8 September, 1918, I Corps Historical Files, Entry 990-32.1, RG 120, NA.

31. Field Order No. 49, HQ I Corps, AEF, 8 September, 1918, I Corps Historical Files, Entry 990-32.1, RG 120, NA.

32. While this tactic had been used before by the Allies, most notably by the British at Cambrai in November 1917, it appears that the AEF was most heavily influenced by the German Army's more dramatic and repetitious implementation of this tactic in their five spring offensives in 1918.

33. Field Order No. 49, HQ I Corps, AEF, 8 September, 1918, I Corps Historical Files, Entry 990-3.1, RG 120, NA; Captain Erwin Davis, "Operations Report and History of the 2nd FA Brigade," 2nd FA Brigade Historical File, Entry 1246-11.4, RG 120, NA.

34. Field Order No. 49, and "Plan for Use of Artillery (Modified)," by General Vincent (French Army), Commander of Artillery, U.S. I Corps, I Corps Historical Files, Entry 990-60.2, RG 120, NA.

35. Operations Order No. 1, HQ 2nd FA Brigade, 9 September 1918, 2nd FA Brigade Historical File, Entry 1246-32.12, RG 120, NA.

36. Spaulding and Wright, *The Second Division*, p. 151.

37. The French guns participating in the barrage were supplied with the 1917 Model "D" shell, which had an eleven-kilometer range, thus enabling them to fire the whole barrage without advancing. Captain Erwin Davis, "Operations Report and History of the 2nd FA Brigade," p. 17, 2nd FA Brigade Historical File, Entry 1246-11.4, RG 120, NA; Operations Order No. 1, HQ 2d FA Brigade, 9 September 1918, 2nd FA Brigade Historical File, Entry 1246-32.12, RG 120, NA.

38. Brigadier General Hanson Ely, "Infantry in Recent Operations," p. 5, lecture on 10 March 1919, to ACAS, in Folder 22, ACAS Historical Files, Entry 371, RG 120, NA.

39. "Report of Operations, Sept. 12–17, 1918, 17th FA," Folder 18, 2nd Division files, G-3 Reports, AEF/GHQ, Entry 270, RG 120, NA; "Report of Operations, Second Division, 12–16 Sept, 1918," *RSD*, vol. 6; Captain John D. Clark, "My War," p. 23, unpublished memoir in 15th FA folders, 2nd Division boxes, World War I Manuscript Collection at USAMHI.

40. "Report of Operations, 2nd Battery, Trench Artillery," 2nd Battery file, Organizational Records, Trench Mortar Batteries, Entry 1243, RG 120, NA.

41. "Report on Operations, 2nd FA Brigade, between Sept 12th–15th," 2nd FA Brigade Historical File, Entry 1246-33.0, RG 120, NA; John A. Lejeune, *Reminiscences* p. 315.

42. Dastrup, *King of Battle,* p. 170; "Report on Operations, 2nd FA Brigade, between Sep. 12th and 15th," 2nd FA Brigade Historical File, Entry 1246-33.0, RG 120, NA.

43. Lanza, "Artillery Support," p. 66; Captain Erwin Davis, "Operations Report and History of the 2nd FA Brigade," p. 17, 2nd FA Brigade Historical File, Entry 1246-11.4, RG 120, NA.

44. "Report of Operations, Second Division, 12–16 September, 1918," *RSD,* vol. 6; "Report of Operations, 15th FA, 17 September, 1918," Folder 17, 2nd Division files, G-3 Reports, AEF/GHQ, Entry 270, RG 120, NA.

45. The 3rd Brigade, whose two regiments served as the assaulting units during the entire attack of 12 September, reported just 369 total casualties for the whole operation. "Report of Operations, Third Brigade, Sept 17th, 1918," *RSD,* vol. 7.

46. "Report of Operations, Sept 12–15, 1918, 23rd Infantry," Folder 73, 2nd Division files, G-3 Reports, AEF/GHQ, Entry 270, RG 120, NA; "Report of Operations of Commanding General, 4th Brigade, Sept. 10–18, 1918," *RSD,* vol. 7; "Report of Operations of the Ninth Infantry, USA, from Sept 12th to 15th, 1918," Folder 63, 2nd Division files, G-3 Reports, AEF/GHQ, Entry 270, RG 120, NA; "Report of Operations, Third Brigade, Sept 17th, 1918," *RSD,* vol. 7.

47. "Attachment to Report on Operations, 2nd FA Brigade, Sept. 12–15," Folder 14, 2nd Division files, G-3 Reports, AEF/GHQ, Entry 270, RG 120, NA.

48. "Report of Operations, Sept 12–15, 1918, 23rd Infantry," Folder 73, 2nd Division files, G-3 Reports, AEF/GHQ, Entry 270, RG 120, NA; "Report of Operations of Commanding General, 4th Brigade, Sept. 10–18, 1918," *RSD,* vol. 7.

49. "Report of Operations of the Ninth Infantry, USA, from Sept 12th to 15th, 1918," Folder 63, 2nd Division files, G-3 Reports, AEF/GHQ, Entry 270, RG 120, NA; "Report of Operations, Sept 12–15, 1918, 23rd Infantry," Folder 73, 2nd Division files, G-3 Reports, AEF/GHQ, Entry 270, RG 120, NA.

50. The 17th FA moved just one of its three battalions (eight howitzers) forward on the first day, while "all vehicles and teams of the other two battalions were used for transporting ammunition up to this battalion from the advance ammunition dump." "Report of Operations, Sept 12–17, 1918, 17th FA," Folder 18, 2nd Division files, G-3 Reports, AEF/GHQ, Entry 270, RG 120, NA.

51. General Ely testified to the difficulty of the terrain in a postwar lecture, admitting that there were thirteen successive trench networks behind the German front-line position, some of which were protected by sections of barbed wire that were between fifty and seventy-five meters deep. The 17th FA reported that despite maintaining "direct communication" with division and artillery brigade headquarters and having a liaison team with the attacking infantry brigade, "it was frequently difficult, after the first day, to know just where our advanced infantry were." Brigadier General Hanson Ely, "Infantry in Recent Operations," p. 6, lecture on 10 March 1919, to ACAS, in Folder 22, ACAS Historical Files, Entry 371, RG 120, NA; "Report on Operations, 2nd FA Brigade, between

Sep. 12th and 15th," 2nd FA Brigade Historical File, Entry 1246-33.0, RG 120, NA; "Report of Operations, 15th FA, 17 September, 1918," Folder 17, 2nd Division files, G-3 Reports, AEF/GHQ, Entry 270, RG 120, NA; "Report of Operations, Sept 12–17, 1918, 17th FA," Folder 18, 2nd Division files, G-3 Reports, AEF/GHQ, Entry 270, RG 120, NA; and "Report of Operations, 2nd Division, 12–16 September, 1918," *RSD*, vol. 6.

52. Lejeune, *Reminiscences*, p. 332. The 15th FA reported that it provided just one hour's worth of a standing barrage from 13 to 14 September, and General Bowley stated that "no operation of importance was executed by the artillery upon the completion of the first attack." See "Report on Operations, 2nd FA Brigade, between Sep. 12th and 15th," 2nd FA Brigade Historical File, Entry 1246-33.0, RG 120, NA; "Report of Operations, 15th FA, 17 September, 1918," Folder 17, 2nd Division files, G-3 Reports, AEF/GHQ, Entry 270, RG 120, NA.

53. Pershing, *My Experiences*, II: 270. Rod Paschall asserts that Pershing's casualty figure of 7,000 must have included only AEF casualties, excluding the French casualties, and only those AEF casualties sustained on the first two days of the attack, in which the salient was actually reduced. The ABMC lists just 8,512 total casualties for all of the AEF divisions that saw action at St. Mihiel from 12 to 16 September. Only Edward Coffman offers a figure of the number of German casualties, besides the number of prisoners; he claims that only about 2,300 Germans were killed or wounded. Coffman, *War to End All Wars*, p. 283; see Paschall, *Defeat of Imperial Germany*, p. 181; ABMC, *American Armies*, pp. 165–66.

54. The ABMC reports total 2nd Division casualties at a slightly lower number, 1,477. *American Armies*, p. 165. The figure used in the text is taken from Report of 2nd Division Commander (Major General John A. Lejeune) to Assistant Chief of Staff, G-3, GHQ, AEF, Subject: "Operations Information," 30 December 1918, in Reports and Other Records Regarding the History of AEF Units, Divisional Skeleton Histories, Entry 443, RG 120, NA.

55. Many of the pertinent German Army documents relating to the defense of the St. Mihiel salient are included in the American official history. See *USAWW*, 8: 288–324.

56. *USAWW*, 1: 42.

57. "Report of Operations, Third Brigade, Sept 17th, 1918," *RSD*, vol. 7.

58. Major General Charles P. Summerall, "Comments by the Corps Commander Upon the Operations of the Fifth Army Corps," in V Corps Historical File, Entry 1118, RG 120, NA.

59. Smythe, *Pershing*, p. 187.

60. Pershing, *My Experiences*, II: 270. Brigadier General Douglas MacArthur, a brigade commander in the 42nd Division, slipped behind enemy lines on the night of 13 September and viewed the town of Metz with binoculars. He returned to his post convinced that Metz was "practically defenseless" and said that if his brigade was authorized to continue the attack, he would be in Metz within a day. Smythe, *Pershing*, p. 188.

5

Blanc Mont: The Set-Piece Attack vs. the Breakthrough

. . . the greatest single achievement of the 1918 campaign.[1]
General Henri-Philippe Pétain
Commander-in-Chief of the French Army

Not slated to immediately take part in the First Army's Meuse-Argonne campaign, the 2nd Division left the St. Mihiel sector for a brief rest, and for the first time following a major operation was in relatively good condition. In the words of Captain John D. Clark, of the 15th FA, the fighting at St. Mihiel was "really quite easy compared with our Soissons experience."[2] The division, brigade, and regimental commanders were proud of the performance of their troops in the attack.

Nevertheless, General Bowley undertook to tighten up the problem of liaison between his artillery forces and the infantry units they supported. On 27 September he issued a memorandum that established updated rules and regulations to guide liaison between the two branches. Artillery liaison officers with infantry units were directed to know the exact locations of artillery command posts and batteries, as well as the nearest telephone connections available. They were required to take three couriers with them, whom they themselves could select, and would have at their disposal one motorcycle, one bicycle, and a horse. Liaison officers assigned to the assaulting infantry battalions were authorized double these amounts. The liaison officers were also required to send back the exact locations of critical infantry command posts and, more importantly, of the actual front-line positions themselves. Artillery commanders were directed to send a replacement courier to each liaison officer upon any movement of command post or batteries. They were also admonished that it was "the duty of the 75m/m regimental and battalion commanders to establish telephone communications forward with the infantry they support," instead of waiting for the infantry to contact them. After securing liaison forward, they should then do the same with artillery headquarters in the rear. The liaison officers were also reminded that if they couldn't reach their own unit for "quick support in an emergency," they should call direct to *any* other artillery unit and repeat that process until sufficient support was secured. That was "their primary duty," regardless of which battalion, regiment, or brigade they themselves were from.[3]

The division did not have long to wait to see if the new system would improve coordination between the infantry and the supporting artillery in battle. By the time the memorandum was received by the artillery regiments, the division had been sent by Pershing to serve in the reserve of the French Fourth Army, which was attacking north in the Champagne region, on the immediate left flank of the American offensive in the Meuse-Argonne. The French XXI Corps had attacked with the rest of the Fourth Army on 26 September and had advanced about five kilometers. However, the French attacks ground to a halt after the first couple of days of slow but steady advances. By 30 September, German forces utilizing the naturally strong positions of an elevation called Blanc Mont Ridge had brought the XXI Corps offensive to a complete standstill.[4]

The high point of the ridge, Blanc Mont, was part of the Hindenburg Line, and was the principal observation post for the German forces in the sector.[5] General Henri Gouraud, commander of the French Fourth Army, stated that "possession of Blanc Mont was essential to the Germans if they wished to maintain their line in the Champagne." Unfortunately for Gouraud, the German forces had taken the strongest position of the ridge and fortified it so heavily that many French officers believed it would never be taken from the front. On 27 September Gouraud confided to Lejeune that his French divisions were "worn out from the long strain of continuous fighting . . . and the heavy casualties they have suffered," and that it was "doubtful" that they were "equal to accomplishing this difficult task."[6]

Lejeune immediately responded with a level of confidence that would have made Pershing proud. He assured Gouraud that if the 2nd Division attacked on a narrow front, it would "take Blanc Mont Ridge, advance beyond it, and hold its positions there."[7] Fresh from the success of St. Mihiel, Lejeune played the part of the supremely confident AEF commander, sure that his unit could accomplish the task (though it is doubtful he had even seen the terrain except on a map), without any hint of a demand for additional time to prepare, special assistance from corps and army level artillery, aviation, or armored forces, or even supporting attacks on his flanks. While the neglect of these other factors demonstrate a classic "Pershing" attitude of aggressiveness, Lejeune's statement actually implied that his unit would accomplish this mission as a limited, set-piece operation, and not as the first stage of an unlimited advance through the enemy lines and into "the open." It appears that he might have had some understanding of the strengths and limitations of his unit.

On 28 September Gouraud told Lejeune that the French High Command had approved the attack by the 2nd Division on Blanc Mont Ridge. The division spent the next two days moving toward the front, and on the night of 1 October its infantry units moved into the front line trenches (some of which had to be retaken from the enemy), with orders to make the attack on the morning of 3 October as part of the French XXI Corps.[8] On 2 October the division's senior commanders met with General Naulin, the XXI Corps commander, and the plan was altered to make use of a small bulge in the line just east of the divisional front. The new plan of attack was probably the most complex used by the division throughout the war (see Map 5.1). The two infantry brigades were to attack simultaneously, each with a frontage of about two kilometers, but the 3rd was to make a converging attack on the eastern edge of the ridge from a starting position nearly two kilometers to the east of the 4th Marine Brigade. As the Marines attacked the western portion of the ridge, containing Blanc Mont, the two brigades would cut off a triangle-shaped section of enemy territory which would be cleaned up by supporting battalions of each brigade.[9]

The new plan meant that the 3rd Brigade would have to make a difficult night march under constant enemy shelling. Furthermore, while the brigade and regimental commanders could begin making dispositions for the new plan based on verbal orders, they would have to wait until the official corps orders were received for confirmation of some details. The corps order was not received by division headquarters until after 10:00 PM on 2 October. When the corps orders arrived, they had to be translated, analyzed, and put into the official division orders that would be sent to the attacking units. Consequently, some attacking units did not receive the attack orders until just prior to "H" hour.[10]

The corps orders announced that the 2nd Division would be supported by a general advance of the whole French Fourth Army. On the division's immediate right would be the French 67th Division and on its left would be the French 21st Division, which was part of the French XI Corps. While the attack was focused on the limited objective of taking Blanc Mont Ridge, all divisions were ordered "to not worry about the alignment. If a unit is delayed its neighbors are to continue to advance

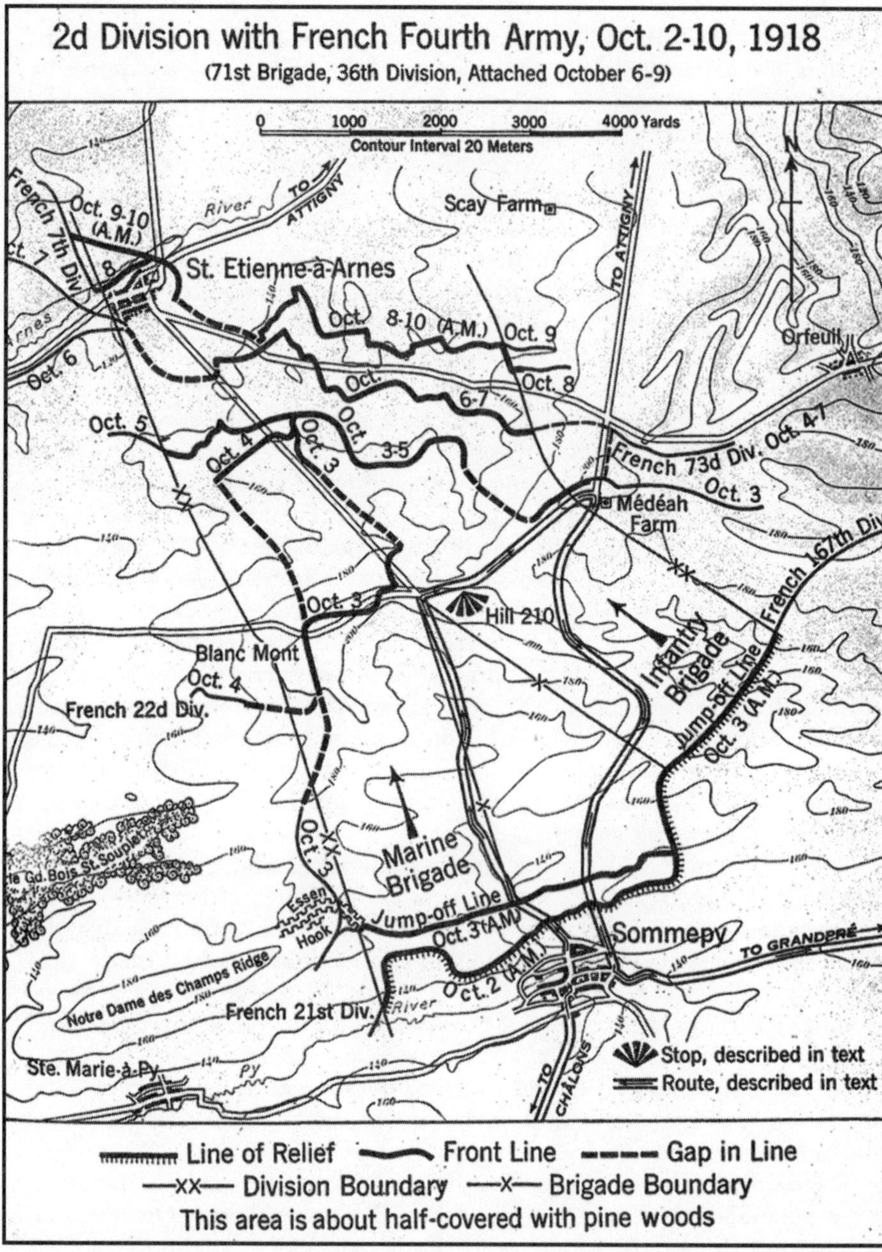

Map 5.1. 2nd Division operations in the Blanc Mont sector with the French Fourth Army, 2–10 October 1918. Notice the initial separation between the infantry and marine brigades, as well as the severe salient they held after outstripping the French divisions on their flanks by the night of 3 October. Taken from ABMC, *American Armies and Battlefields in Europe*, p. 349.

only taking care to cover their flanks." This was an aggressive and potentially dangerous provision for an attack with supposedly limited objectives, and it was to have a considerable impact on the 2nd Division.

The 2nd Division was augmented with two French regiments of 75s, two French battalions of 155mm howitzers, and two French battalions of 120mm long guns. However, while this force approximated the amount of artillery support given to the division for its St. Mihiel attack, these guns came with an additional duty. Unlike at St. Mihiel, it became the task of the divisional artillery complement to "take apart . . . all batteries" in the 2nd Division's zone. The divisional artillery, assisted by the staff of one French artillery regiment, was to "assure the preventative neutralization of batteries . . . which might interfere with its progress by flank fire."[11] This was a tall order, and one that the divisional artillery was not trained, equipped, or experienced enough to accomplish. As additional support to the infantry, each brigade was given twenty-four light tanks.

There was to be no heavy preliminary bombardment; in fact, the artillery instructions called for fire to progressively diminish in intensity throughout the night. It was to abate entirely during the last hour preceding the assault, with only harassing fire permitted during that time. Finally, five minutes prior to the attack, a "brutal and intense preparation" was called for "to force the enemy to the ground."[12] During this time, all guns in the 2nd Division were to fire at a maximum rate, with the 75s placing a raking fire of high explosive and shrapnel shells over the entire enemy area, and the 155s concentrating on strong points, trenches, machine-gun nests, command posts, and telephone stations.[13]

Immediately at "H" hour the divisional artillery was to put down a rolling barrage in each brigade sector, advancing 100 meters every four minutes, which was to take the 3rd and 4th Brigades three and four kilometers, respectively, into the German positions. The angular approach of the 3rd Brigade complicated the rolling barrage in its sector, as the supporting artillery could not be positioned directly behind the brigade's lines. Lejeune's order directed that

> troops of the right (east) brigade will be cautioned not to follow the barrage too closely because the guns with which this barrage is made will not be fired in the axis of the advance, but at an angle thereto. Upon reaching the Division Objective, troops of the right (east) brigade will be held strictly within that part of the objective assigned to that brigade, until the arrival of the left (west) brigade.[14]

Just beyond the first objective, which the 3rd Brigade was scheduled to reach first, a standing barrage was to be put down until thirty minutes after the Marines arrived on the same line. The barrage would then roll forward again at the same rate for another 1,300 meters to allow the leading regiments to send out their forward patrols, and then hold there as a standing barrage to guard the gains of the advance.[15]

While the 75s fired the rolling barrage, the 155s were to fire in concentrations on strong points in advance of the barrage, lifting to deeper targets whenever the rolling barrage came within 400 meters of the current target area. Approximately a third of all the 155s were to fire solely on the enemy trenches along Blanc Mont Ridge, the attack's main objective. Two French battalions of 75s were designated to shoot only smoke shells in advance of the rolling barrage, in an attempt to screen the attacking infantry. As at St. Mihiel, the 12th and 15th artillery regiments each kept one battalion hitched and ready to advance at the earliest possible moment.[16]

Promptly at "H" hour, 5:50 AM, 3 October, the assault troops from both brigades jumped out of their trenches, threw out their advance skirmishers, and moved forward behind the rolling barrage. The leading battalions of the 3rd Brigade were met shortly by a strong counterbarrage of high-explosive shells which came "smashing down . . . among all the troops." General Ely, who was far enough forward to witness the assault, later said that if the Germans had shot just sixty percent shrapnel instead of all high explosive, there would have been "tremendous casualties."[17] Clearly, the counterbattery effort had not been entirely successful. Nevertheless, the 3rd Brigade advanced through this heavy artillery fire, was able to maintain contact with its own rolling barrage, and decimated the German regiment that stood between it and the ridge. By 8:40 AM the brigade had taken its objective, "without serious loss" in casualties, and had maintained good contact with the Marines on its left and with the French 67th Division on its right. One battalion was quickly pushed forward to establish and maintain an outpost position, and Ely reported that his unit was able to continue the advance if necessary.[18]

The 4th Brigade did not have as much trouble with enemy artillery fire as it did with fire from German machine-gun nests, the most trouble coming from its immediate left flank. While it advanced steadily behind the rolling barrage of the 12th FA, the French 21st Division on its left made absolutely no progress against a formidable German trench system in front of the base of the ridge called the "Essen Hook." As a result, the left flank of each successive battalion in the 4th Brigade was somewhat retarded by the fire, and a detachment from the 5th Regiment was sent to the flank with 37mm "one-pounders" to help clear out "the Hook."[19] To the front, though, the leading battalions continued without much difficulty, at one point even shooting a flare to lengthen the barrage ahead of schedule. By 8:30 AM the leading elements had reached the objective, reported that "losses were light," and "began the work of consolidating the position and establishing the line of outposts to the front."[20]

The first part of Lejeune's promise to General Gouraud was made a reality—the division had indeed taken Blanc Mont Ridge, and despite the pessimistic French appraisals of the German position, had done so with comparatively light casualties. The set-piece attack was a success. As at St. Mihiel, the infantry commanders heaped praise on their supporting artillery. General Ely said that the 15th FA "did excellent work" and that "close liaison was established and maintained . . . infantry and artillery worked well together." Immediately after taking the objective, General Neville, the commander of the 4th Brigade, reported that "the barrage was beautiful." He later confirmed this by stating that his Marines "testified to the excellency of the rolling barrage and the accuracy of special fire."[21] What remained to be seen was how well the 2nd Division could do in the other two parts of Lejeune's bold promise, moving beyond and holding the ridge.

At 9:00 AM Lejeune reported to the French Fourth Army that his artillery was already advancing and would be "all well forward by this afternoon." He also reported that his gunners were not receiving enough ammunition from the French supply officers, and warned that he would need an increased allotment "if we are going to exploit and carry another attack this afternoon." By 12:30 PM, at least one battalion of the 15th FA was in a forward position, north of the town of Sommepy, while the other light battalions were leapfrogging their way toward that position, ensuring constant fire support to the infantry. During the course of the day, the 17th FA and French 120mm guns advanced to the same forward location. However, much of the other French artillery that had augmented the firepower of the division was being taken from the division and given to struggling divisions elsewhere along the front. Nine light and three heavy batteries from the French 61st Division that had helped with the morning attack were pulled from the 2nd Division and assigned to the French XI Corps, which was making no headway at all on the left.[22]

While Lejeune may have liked to have held on to that additional artillery, he knew where it was needed most. From 9:00 AM on, Lejeune had made constant pleas to both the XXI Corps and the French Fourth Army to get him some support on his left flank. At one point he even questioned the courage of the flagging French 21 Division, asserting that "we don't think they attack with as much spirit as possible." Anticipating that the French army and corps commands would push for him to exploit the success of his morning attack, Lejeune sent a preemptive warning to them that his division could not advance further until his left flank was protected. General Naulin immediately ordered the 2nd Division to leave a battalion on its flank and continue the advance. He then tried just about everything he could think of to solve Lejeune's problem. He sent unrealistic assurances that the 21st Division would eventually come forward. He said he "directed" that the artillery of the French XI Corps was to protect the flank with fire. Eventually he ordered a tired reserve division, the French 170th Division, to march into the left flank. Despite the corps commander's efforts, hours passed without any effective support arriving on the left.[23]

Nevertheless, in compliance with corps directives, at 2:00 PM Lejeune issued an order to continue the attack. After announcing that the morning attack was "a complete success," he ordered his brigades to continue the advance a couple more kilometers to the northwest, where "a position of resistance will be established and held," and outposts would be pushed forward yet another kilometer. The fire-support plan for the afternoon attack was given in one simple sentence, "the Second Field Artillery Brigade will support the movement."[24] Obviously, the details were to be worked out by others. It was a far cry from the more elaborate artillery fire plan that was included in the orders for the morning attack.

As the front-line troops began to receive more and more enemy artillery fire, the details of the afternoon attack plan were sent back and forth between the infantry commands and their supporting artillery.[25] At 3:00 PM the word went out to the 3rd Brigade that the afternoon attack would begin at 4:00 PM, with a five-minute artillery preparation. At 4:05 the rolling barrage would begin at the point where it had stopped that morning, and roll forward at a rate of 100 meters every four minutes.[26] However, the Marine brigade did not get the attack orders until later in the day, and due to the problems they were having on their left flank, decided to delay the attack until the next morning.

On the right, the attack of the 3rd Brigade went reasonably well, its leading infantry units advancing nearly another two kilometers towards the town of St. Etienne. Unfortunately, they were the only infantry to move forward along the whole front line of the sector, and at nightfall the 2nd Division held a dangerous salient. To the left, the French 21st Division still had not even retaken the Essen trench, exposing the Marines to increasingly heavy machine-gun and artillery fire from the flank and rear. To make matters worse, the French 167th Division on the right of the 3rd Brigade, which had kept pace in the morning attack, made no progress at all in its afternoon assault. The 3rd Brigade received heavy machine-gun and artillery fire from its right and left flanks, and even from the right and left rear, all night long.[27] Early the next morning, 4 October, the Marines complied with the orders of the previous day, advanced another two kilometers, and came up on the left of the 3rd Brigade.

The division had successfully advanced beyond the ridge, as Lejeune had said it would, and the division commander was intent on consolidating his gains. However, the French army and corps commanders were beginning to receive instructions directly from the Allied generalissimo,

Marshal Foch. The French XXI Corps announced that "General Foch has just learned of the success of the XXI Corps and of the Second Division, U.S., which is assigned thereto. He orders that this success be exploited to the limit. All must press forward at once, without hesitation. The breach is made; the enemy must not be given time to repair it."[28]

And so ended any idea of a limited attack, in which the 2nd Division would be allowed to thoroughly reorganize its position and adequately prepare for additional advances. The French Army had its own officers who shared Pershing's love of the unlimited, unrelenting offensive, and Foch was their moral leader. However, whereas the American combat experience was just starting to make American officers doubt the effectiveness of this "cult of the offensive" and the doctrine it fostered, the French Army had developed a counterdoctrine, with its own apostle—General Henri-Philippe Pétain. Having witnessed the collapse of the French Army in 1917, caused in large part by the execution of the doctrine of the unrelenting offensive, Pétain had developed a more conservative approach that "insisted on the massive use of firepower and carefully orchestrated assault."[29] While the different doctrines competed with one another in the French Army, and though Pétain was the commander-in-chief of that army, in this instance the demands of Foch won out and the remainder of the fighting near Blanc Mont was colored by his insistence on continual offensive pressure, regardless of the specific requirements of any local engagement.

On the morning of 4 October, Lejeune issued another attack order, calling for the resumption of the offensive toward the town of St. Etienne. However, the order did not fix the time for the attack because Lejeune was waiting for the French divisions on his flanks to make some move forward. He later said that the corps orders directing his division to drive an ever-deeper salient in the enemy lines was "courting certain destruction."[30] However, before noon the division received reports that the neighboring French divisions had advanced, and Lejeune set the time for the day's attack at 2:30 PM. The attack was attempted but was "held up by the action on our front and flanks, of hostile artillery and machine-gun fire and particularly, from a large German strong point on the west end of Blanc Mont." After suffering heavy losses, the infantry had to return to its starting positions. Clearly, the division's ability to engage in full-fledged, penetrating attacks had decreased.[31]

One problem was that the French divisions never came up on the flanks. While the French 22nd Division, which had just replaced the 21st Division, advanced quickly on the left, it veered to the west, away from Blanc Mont and the German strong point on its western slope. The situation was worse on the right, where the French 73rd Division was unable to advance, even though it too was "fresh," having just replaced the 167th Division. Lejeune decided that his division should attempt no further advances until the strong German position on the west edge of Blanc Mont (to his left and rear) was completely eliminated. After a hurried and poorly supported attempt to reduce the position that afternoon, the Marines and the 12th and 17th FA Regiments coordinated thorough plans to systematically reduce the position. The gunners kept up harassing fire on the enemy position throughout the night. Early the next morning, after an intense sixty-minute bombardment of the position in which 800 shells were fired on the enemy, a battalion of Marines attacked behind a heavy rolling barrage and took the strong point. The attack was so well coordinated that the Marines captured the German detachment of nearly 300 men and over 80 machine guns without a single American casualty.[32]

Despite the failures in the general advance of the previous day, General Naulin ordered that the advance was to continue, and in compliance Lejeune issued orders for another attack on 5 October, with the primary advance to be made by the 6th Marine Regiment.[33] As with the attempt on the previous day, the attackers met heavy resistance from nests of machine guns, this time in their front. Although they were able to make some headway, the losses were again severe. Therefore, in a return to the strategy of the previous day, the commander of the 6th Regiment, Colonel Harry Lee, "perfected plans" for a morning assault that were "elaborate and thorough." All three regiments of artillery were to concentrate their fire on the enemy positions for an hour-long bombardment, which would be followed by a rolling barrage. Coordinated attacks by the leading battalions in each brigade were programmed.[34] The attack was made at 6:30 AM, 6 October, and was over by 9:30 AM with "the objective obtained, consolidated and strengthened," and outposts sent to the front.[35] Immediately after seizing the enemy position, some front-line Marines identified an imminent counterattack, which was "broken up by a special barrage which was laid within eight minutes after receipt of request."[36] Again, the smaller, scripted, set-piece attack succeeded where the hurried, general offensive had failed.

On the night of 6 October all but two front-line battalions of the 2nd Division were relieved by units of the 71st Brigade of the U.S. 36th Division, which had not seen front-line duty of any sort until that date. Since the 36th Division did not have its own artillery, the 2nd FA Brigade remained in position and supported the attacks of the 71st Brigade. The 71st Brigade, with some assistance from the 3rd and 4th Brigades, continued to attempt general attacks on 9, 10, and 11 October, at which time the Germans to their front began a twenty-four kilometer withdrawal to a position behind the Aisne River. The 36th Division, with the 2nd FA Brigade supporting, pursued them the entire distance, arriving on the Aisne on 13 October.[37]

The fighting at Blanc Mont presented a clear indication of the strengths and weaknesses of the 2nd Division. Its planning and execution of the very challenging attack on 3 October was exceptional. General Gouraud said the division's first-day advance of nearly six kilometers against a determined enemy in a formidable defensive position was "remarkable." Unlike at St. Mihiel, the German forces on Blanc Mont Ridge were not carrying out any kind of a planned withdrawal at the time; in fact, they fought so tenaciously on that ridge for the express purpose of trying to avoid a general withdrawal of all German forces south of the Aisne. Furthermore, the 2nd Division successfully took all its initial and secondary objectives while the French divisions on its flanks failed to continue the advance, or in the case of the French 21st Division on the left, to advance at all. As previously mentioned, the artillery work in the first day's assault was excellent, with the exception of the counterbattery effort, which was a mission that few, if any, World War I divisional artillery contingents were capable of performing with any degree of effectiveness. As for the liaison between the infantry and artillery, which had deteriorated at St. Mihiel, all indications are that it improved markedly. With respect to the support that the 3rd Brigade received from the 15th FA, General Ely reported that "it is not believed that closer cooperation between infantry and artillery could have been obtained."[38]

However, it is clear that after the initial set-piece assault, the division's attacks became less and less effective, and more and more costly. While the division suffered 4,975 casualties from 1–10 October, and advanced nine kilometers, the official reports indicate that the great majority of those casualties were incurred in the general attacks made after the first day.[39] Why was this the case? After the first attack, at the same time that the division was being ordered forward, and with its position becoming increasingly complex, the amount of time it had to plan the general attacks decreased. Similarly, although its salient position increased the length of front that required fire support from its artillery, the number of guns available to the division was continuously reduced. The division became increasingly capable of successfully performing smaller and smaller operations. Unfortunately, this did not mesh with the excessively optimistic plans of certain French commanders. Any thought of the 2nd Division engaging in a series of well-choreographed set-piece attacks died very early on the first day, when the senior French commanders thought they smelled the much dreamed of, but rarely seen, "breakthrough." By noon of the first day, General Naulin had ordered his cavalry to form in the rear of the 2nd Division, so it would be "ready to break through when the opportunity presents itself."[40] The cavalry was never needed; no breakthrough ever occurred. But from then on, the division was ordered forward every day, even though its flanks were dangerously exposed, its offensive firepower greatly reduced, and its infantry regiments were becoming weaker and more tired.

It is interesting to note that at the time when the French army and corps commanders (at Marshal Foch's insistence) were prodding the 2nd Division to continue its general forward attacks, General Pershing was giving the same directions to his divisions that were grinding forward in the American First Army's Meuse-Argonne campaign. After the First Army surged forward as far as ten kilometers in spots on the first day's attack, the offensive slowed down, and casualty figures began to rise. While aware of the logistical problems that were contributing to the slow advance, Pershing continued to stress the need for constant forward pressure. He ordered:

> It is of extreme importance that the American First Army drive forward with all possible force. There is evidence that the enemy is retiring from our own front. Our success must be followed up with the utmost energy . . . to prevent the enemy from forming his shattered forces. I am counting on the splendid spirit, dash and courage of our army to overcome all opposition.[41]

After the initial assault, which was well supported by massive firepower and detailed planning, Pershing, like Foch and the other senior officers who shared a love of the unlimited offensive, was reduced to demanding that his attacking troops allow moral qualities to replace the firepower and organization that had been so influential in the first attacks. Though it may seem odd that Pershing made no appeal to the experience and skill of his forces, he knew exactly what he was saying. He later commented, "it was thought reasonable to count on the vigor and the aggressive spirit of our troops to make up in a measure for their experience, but at the same time the fact was not overlooked that the lack of technical skill might considerably reduce the chances of complete success against well organized resistance of experienced defenders."[42] Obviously, for Pershing, the hope of breaking into some form of "open warfare" was worth the risks. However, what those steeped in the "cult of the offensive" apparently did not understand was that there were more effective ways to aggressively employ those troops, but they hinged on matching the style and scale of the operations to the strengths and capabilities of the soldiers, and not on any abstract doctrine or traditional view of warfare.

While the 2nd Division fully accomplished all it had set out to do at Blanc Mont, it showed signs of the brutal fighting it had experienced. Four days after leaving the front lines, AEF inspectors reported that the division required over 2,000 soldiers and some 1,600 Marines to refill its four infantry regiments, and nearly 300 artillerymen to fill holes in the three artillery regiments. It also needed an immediate issue of 500 horses and mules to regain its transportation capability. Although replacements were becoming harder and harder to find in the AEF, the 2nd Division received all the troops it needed, as it had been recalled to the American First Army to take place in one more great offensive—the third phase of the Meuse-Argonne campaign, scheduled to begin on 1 November.[43]

NOTES

1. Quoted in Coffman, *War to End All Wars*, p. 284.

2. Captain John D. Clark, Diary, 23 September 1918, in 15th FA folders, 2nd Division boxes, World War I Manuscript Collection at USAMHI.

3. Memorandum, "Liaison Between Infantry and Artillery: General System of Artillery Support," from HQ 2nd FA Brigade, 27 September, 1918, 2nd Division Historical File, Entry 1241-32.11, RG 120, NA.

4. By comparison, the American First Army had by this date advanced approximately ten kilometers. ABMC, *American Armies*, p. 333.

5. U.S. Army, General Staff, War Plans Division, Historical Branch, "Blanc Mont (Meuse-Argonne-Champagne)," Monograph No. 9, April 1920, at USAMHI.

6. Report of General Gouraud to Marshal Foch, quoted in "Notes on Blanc Mont," 2nd Division Historical File, 1241-33.2, RG 120, NA; Lejeune, *Reminiscences*, p. 342.

7. Lejeune, *Reminiscences*, p. 342.

8. The attack was originally planned for 2 October, but was delayed upon the suggestion of Lejeune for the following reasons: his artillery batteries were not going to be in position in time to support an attack on that date, he wanted to give his commanders the chance to see the terrain in the daylight, and the jump-off trenches needed to be cleared of German outposts. Lejeune, *Reminiscences*, p. 345.

9. General Operation Order, No. 3703/3, XXI Army Corps, Staff, 3rd Bureau, 2 October 1918, 2nd Division Historical File, Entry 1241-32.5-33.0, RG 120, NA; Field Order No. 35, HQ 2nd Division, 2 October 1918, Folder 5, 2nd Division files, G-3 Reports, AEF/GHQ, Entry 270, RG 120, NA.

10. The commander of the leading battalion of the 6th Marines received his copy of the orders exactly at "H" hour, and did not even get a chance to read them until he reached his objective. "Operations Report, 4th Brigade, Marines, October 1st–10th, 1918," *RSD*, vol. 7.

11. Document No. 7538/3, Fourth Army, XXI Corps, Artillery Command, Staff, 2 October 1918, 2nd Division Historical File, Entry 1241-32.7, RG 120, NA; Captain Erwin Davis, "Operations Report and History of the 2nd FA Brigade," p. 20, 2nd FA Brigade Historical File, Entry 1246-11.4, RG 120, NA.

12. General Operation Order, No. 3703/3, XXI Army Corps, Staff, 3rd Bureau, 2 October 1918, 2nd Division Historical File, Entry 1241-32.5-33.0, RG 120, NA.

13. Captain Erwin Davis, "Operations Report and History of the 2nd FA Brigade," p. 21, 2nd FA Brigade Historical File, Entry 1246-11.4, RG 120, NA.

14. Field Order No. 35, HQ 2nd Division, 2 October 1918, Folder 5, 2nd Division files, G-3 Reports, AEF/GHQ, Entry 270, RG 120, NA.

15. Field Order No. 35, HQ 2nd Division, 2 October 1918, Folder 5, 2nd Division files, G-3 Reports, AEF/GHQ, Entry 270, RG 120, NA.

16. Captain Erwin Davis, "Operations Report and History of the 2nd FA Brigade," p. 21, 2nd FA Brigade Historical File, Entry 1246-11.4, RG 120, NA.

17. Brigadier General Hanson Ely, "Infantry in Recent Operations," p. 7, lecture on 10 March 1919, to ACAS, in Folder 22, ACAS Historical File, Entry 371, RG 120, NA.

18. The German 410th Infantry reported that it suffered 810 casualties on 3 October, including the death of its commanding officer. War Diaries, 203d Division and 410th Infantry, Oct. 1–10, as cited in Spaulding and Wright, *The Second Division*, pp. 173–74; "Messages Sent and Received (and Incidents)—Blanc Mont," Folder 48, 2nd Division files, G-3 Reports, AEF/GHQ, Entry 270, RG 120, NA.

19. This detachment did excellent work on the Essen Hook, taking possession of the trench system and seizing 100 prisoners. They turned the trench over to the French in that sector and resumed the advance with the rest of the 5th Regiment. However, the French did not occupy the Essen trench with sufficient strength, and the Germans retook the position in force that afternoon. Memo, "Short Account of the cleaning out of the 'Essen Hook,' " 5th Regiment, Marines, in Folder 108, 2nd Division files, G-3 Reports, AEF/GHQ, Entry 270, RG 120, NA.

20. "Operations Report, 4th Brigade, Marines, October 1st–10th, 1918," *RSD*, vol. 7; "Messages Received During the Battle of Blanc Mont Ridge," Folder 5, 2nd Division files, G-3 Reports, AEF/GHQ, Entry 270, RG 120, NA.

21. "Report of Operations, Oct. 1–8," 3rd Brigade and "Operations Report, 4th Brigade" included in "Special Report: Operations of the Second Division (Attack on Massif of Blanc Mont Ridge and the Action in Connection Therewith), Oct. 3–Oct. 10, 1918," p. 17, in G-3 Reports, AEF/GHQ, Entry 270, RG 120, NA; "Messages Received During the Battle of Blanc Mont Ridge," Folder 5, 2nd Division files, G-3 Reports, AEF/GHQ, Entry 270, RG 120, NA.

22. "Messages Received During the Battle of Blanc Mont Ridge," Folder 5, 2nd Division files, G-3 Reports, AEF/GHQ, Entry 270, RG 120, NA; Spaulding and Wright, *The Second Division*, p. 177.

23. The worn-out French 170th Division eventually reported to General Lejeune at 7:00 PM, 3 October, and helped guard that flank. Lejeune, *Reminiscences*, p. 353; "Messages Sent and Received (and Incidents)—Blanc Mont," Folder 48, 2nd Division files, G-3 Reports, AEF/GHQ, Entry 270, RG 120, NA; "Messages Received During the Battle of Blanc Mont Ridge," Folder 5, 2nd Division files, G-3 Reports, AEF/GHQ, Entry 270, RG 120, NA.

24. Field Order No. 36, 2:00 PM, 3 October, Folder 5, 2nd Division files, G-3 Reports, AEF/GHQ, Entry 270, RG 120, NA.

25. At 1:25 PM the 3rd Brigade reported that "the whole front line is being heavily shelled. Request counter-battery work." It is more evidence that the counterbattery effort was weak. See "Messages Sent and Received (and Incidents)—Blanc Mont," Folder 48, 2nd Division files, G-3 Reports, AEF/GHQ, Entry 270, RG 120, NA.

26. "Messages Sent and Received (and Incidents)—Blanc Mont," Folder 48, 2nd Division files, G-3 Reports, AEF/GHQ, Entry 270, RG 120, NA; ABMC, *American Armies*, p. 352.

27. ABMC, *American Armies*, pp. 352–53; Captain Erwin Davis, "Operations Report and History of the 2nd FA Brigade," pp. 21–22, 2nd FA Brigade Historical File, Entry 1246-11.4, RG 120, NA.

28. Document No. 3715/3, 3 October 1918, XXI Army Corps, Staff, Third Bureau in 2nd Division Historical File, Entry 1241-32.7-33.0, RG 120, NA; Spaulding and Wright, *The Second Division*, p. 179.

29. Paschall, *Defeat of Imperial Germany*, p. 225.

30. By the morning of 4 October, both French divisions on the flanks were replaced, the 21st Division by the French 22nd Division and the 167th Division by the French 73rd Division. Spaulding and Wright, *The Second Division*, p. 180–81; Field Order No. 37, 6:00 AM, 4 October, HQ 2nd Division, Folder 5, 2nd Division files, G-3 Reports, AEF/GHQ, Entry 270, RG 120, NA; Lejeune, *Reminiscences*, p. 353.

31. Spaulding and Wright, *The Second Division*, pp. 180–81; ABMC, *American Armies*, p. 353.

32. Spaulding and Wright, *The Second Division*, p. 181; "Messages Received During the Battle of Blanc Mont Ridge," Folder 5, 2nd Division files, G-3 Reports, AEF/GHQ, Entry 270, RG 120, NA.

33. Field Orders No. 38, 5:00 AM, 5 October 1918, Folder 5, 2nd Division, G-3 Reports, AEF/GHQ, Entry 270, RG 120, NA.

34. "Operations Report, 4th Brigade, Marines, October 1st–10th, 1918," *RSD*, vol. 7; Spaulding and Wright, *The Second Division*, p. 185.

35. "Messages Sent and Received (and Incidents)—Blanc Mont," Folder 48, 2nd Division files, G-3 Reports, AEF/GHQ, Entry 270, RG 120, NA.

36. "Operations Report, 4th Brigade, Marines, October 1st–10th, 1918," *RSD*, vol. 7; Spaulding and Wright, *The Second Division*, p. 185.

37. The U.S. 36th Division had only been in France for two months, had not even done a tour at a quiet sector of the front, and did not have its artillery brigade, engineer regiment, or much of its transportation compliment. ABMC, *American Armies*, pp. 356–57.

38. "Report of Operations, Oct. 1–8," 3rd Brigade, in G-3 Reports, AEF/GHQ, Entry 270, RG 120, NA.

39. Again, the ABMC lists a lower casualty figure of 4,821. The figure in the text is taken from the report of 2nd Division Commander (Major General John A. Lejeune) to Assistant Chief of Staff, G-3, GHQ, AEF, Subject: "Operations Information," 30 December 1918, in Reports and Other Records Regarding the History of AEF Units, Divisional Skeleton Histories, Entry 443, RG 120, NA.

40. Message from XXI Corps commander to 2nd Division, 11:30 AM, 3 October, "Messages Received During the Battle of Blanc Mont Ridge," Folder 5, 2nd Division files, G-3 Reports, AEF/GHQ, Entry 270, RG 120, NA.

41. General Orders, No. 20, 28 September 1918, First Army, AEF, *USAWW*, 9: 144.

42. Pershing, *My Experiences*, II:293.

43. By 27 October the inspectors could report that the divisional combat brigades had nearly been refilled. The strength of the 3rd Brigade was 278 officers and 8,607 men, the Marine Brigade had 205 officers and 7,775 men, and the 2nd FA Brigade numbered 203 officers and 4,882 men. However, the division was still short over 2,000 animals. "Inspection Reports of Armies, Divisions, and Corps, 1917–1919," 2nd Division files, Inspector General, GHQ/Administration Staff, Entry 590, RG 120, NA.

6

Meuse-Argonne: Breakthrough of a Sort

If we are to be economical with our men, we must be prodigal with guns and ammunition.[1]

Major General Charles P. Summerall
Commanding General, V Corps

The first phase of the Meuse-Argonne campaign, which began on 26 September, and to a lesser degree the second phase, which started on 4 October, had both been large-scale demonstrations of the successes that could be gained in high-powered initial assaults. However, those offensives also clearly showed the problems and loss in efficiency that the attacking divisions experienced during subsequent attacks. Considering the tremendous strength of the German positions opposing his army, Pershing's plans for his initial attack in the first phase of the offensive were exceptionally bold. The attack orders directed his forces to advance some sixteen kilometers within two days and break through the enemy's main line of resistance, the Kriemhilde Stellung of the Hindenburg Line, at the far end of the deep two-day advance. While the attack on 26 September began well, once the troops advanced beyond the range of their supporting artillery, the advances slowed, then stopped, and the casualties rose. Finally, on 14 October, after weeks of slugging forward, the Army reached the initial objectives it had set for 27 September.[2]

There were three major problems with the employment of artillery in the first phases of the campaign. The first was that plans called for the infantry to continue its initial attack all the way to distant objectives that were surely going to be beyond the range of the supporting artillery.[3]

Pershing and his staff were counting on their infantry being able to successfully carry those distant objectives with little or no artillery support. The second problem was that the attacking corps and divisions attempted to make such a quick, deep penetration without using one of the most tactically decisive weapons of the German Army's spring offensives—gas. Smothering the defending forces with tons of gases of all types was an integral part of each one of Germany's spring offensives in 1918.[4] Pershing also included a massive gas bombardment in his initial plan for the 26 September attack. Unfortunately, the AEF could not secure all the gas shell it needed to carry out the plan. Additionally, the decision as to when, where, and how to use the gas that was available was rightfully left to the corps and division commanders, and they, due to unfamiliarity with the weapon, an almost inexplicable fear of retaliation, and an inflated expectation of what the infantry could accomplish on its own, did not carry through on Pershing's plan for using gas.[5] A third problem was that the army and corps artillery groups were not used to maximum effect in either counterbattery against the German guns or directly supporting the infantry advance with an additional rolling barrage. Due to the expectations of many high-ranking AEF officers that aggressive infantrymen would be able to make successful attacks without the support of heavy artillery, and the fear that the use of heavy guns in an infantry support role would lead to a rash of friendly fire incidents, the corps and army guns were prohibited from firing on positions close enough to the advancing infantry to be of any practical use to them.[6]

However, from mid-October until the end of the month, the American First Army did little but make preparations for its final big push on 1 November. General Liggett, who replaced Pershing as commander of the First Army, made changes tailored to solve the problems that had plagued the first two phases of the offensive. Meanwhile the 2nd FA Brigade left the 36th Division on the Aisne on 28 October and took up positions behind its infantry brigades in the Meuse-Argonne sector two days later. By then plans for the attack on 1 November were almost finalized. General Bowley left Colonel Joseph R. Davis from the 15th FA in command of the artillery brigade, which was still with the 36th Division near Blanc Mont, and had joined General Lejeune to aid him in the last stages of the planning effort.[7] As early as 25 October, the division had produced a "Tentative Plan of Attack," and the infantry brigade, regimental, and battalion commanders were able to study the plan and reconnoiter the ground.[8]

General Liggett assigned the 2nd Division to General Summerall's V Corps, which held the lines in the center of the American First Army. Unlike the previous attacks in this campaign, the plan of attack for 1 November did not call for all divisions in line to make as deep a penetration as possible. The V Corps was called on to make a nine-kilometer deep penetration to seize the heights of Barricourt Ridge, while the other two corps, the I Corps on the left and the III Corps on the right, were to make supporting attacks only. The 2nd Division would advance through two intermediate objectives, each followed by a thirty-minute rest, before taking the final "First Day" objective. Subsequent attacks on following days by all three corps were to drive the enemy north of the Meuse River.[9]

These plans, and the supporting artillery fire plans, were developed in a series of conferences held at V Corps Headquarters attended by the corps and division commanders, their chiefs of staff, all chiefs of artillery, and the division machine-gun officers.[10] While previous attacks counted on the infantry to fight their way forward, this attack, more than any other AEF or 2nd Division attack of the war, put its faith in massive amounts of well-coordinated firepower.

This was especially true in V Corps, where Summerall, an artilleryman and former artillery brigade commander (see Figure 6.1), ordered that "fire superiority, rather than sheer man power be the driving force of the attack."[11] With the full support of General Liggett, army and corps planners addressed each of the three major artillery problems that had plagued the earlier phases of the campaign. First, the plans called for the supporting artillery to fully cover the attacking troops all the way through the first day's final objectives. Of equal importance was the fact that all of the main lines of enemy resistance scheduled to be assaulted on the first day were within a few kilometers of the jump-off positions and were therefore well within range of all the supporting artillery.[12] Second, unlike the earlier drives, AEF artillery was to subject the enemy to a very heavy gas bombardment, in some places beginning as much as two days before the actual assault. The plan called for the use of gas, specifically persistent mustard gas, in a systemic neutralization of the large German artillery groups on each flank, so as to protect the main thrust of the V Corps from the crushing fire that had previously come from the guns in those positions.[13]

Third, the heavy guns of the army and corps batteries were to be extensively involved in providing accompanying fire for the infantry. In previous attacks the use of heavy guns in the infantry support role had been confined to firing concentrations on known or "suspected" enemy positions, and always well ahead of the infantry they were supposed to be supporting. For the 1 November attack, Liggett removed this restriction and directed the "complete use of artillery." His First Army planners therefore programmed the use of army and corps guns in an infantry support role, allowing them to fire their concentrations on a rolling barrage schedule that kept the fire within 1,000 meters of the advancing infantry throughout their advance.[14]

Figure 6.1. Major General Charles P. Summerall. An artilleryman who commanded an artillery brigade, combat division, and army corps before the war ended, Summerall was one of the AEF's foremost apologists for using massive firepower to help the infantry advance. Courtesy McCormick Research Center.

In V Corps, Summerall divided the fire-support plan into four periods: the time preceding the preliminary bombardment, the preliminary bombardment itself, the covering fire for the infantry's initial assault, and the support of the infantry during "subsequent stages of the advance." The first period covered several days, and included the firing of gas, shrapnel, and high-explosive shells by all calibers on "all enemy organizations, batteries, and routes of communication." During this time, American gunners would make sudden bursts of fire from all guns two or three times a day, so that the defending German troops, who fully expected a continuation of the offensive, would not be sure when the real assault was starting and therefore would be "trained" to respond more slowly when it did come. Additionally, AEF artillerymen would use this period to carry out a carefully scheduled and concealed registration of artillery pieces already in their positions.[15]

The fire plan called for the preliminary bombardment to last only two hours, but it was to be accomplished at the maximum rate of fire for each gun. According to the corps commander, the very short preparation was driven by the amount of available ammunition, "as ammunition could not be procured for a maximum rate of fire during a longer period," and Summerall believed it "better to deliver fire at a maximum rate than to employ slow fire over a longer period."[16] This line of reasoning was in step with much of the most current and effective artillery theories of war, such as short German bombardments of the spring offensives and the "hurricane bombardments" of the BEF. As for targeting, the 75s were to hit the enemy front lines and certain German organizations. The 155s were to hit specific towns, woods, and all known and suspected machine-gun nests. Other corps and army guns were to neutralize enemy batteries.[17] To add to the maelstrom, 255 machine guns would fire coordinated barrages into the German positions opposite the 2nd Division.[18]

One important innovation was made to this preparation. Some infantry attacks in previous phases of the offensive had been met immediately by heavy enemy machine-gun fire. Often this delayed the infantry's advance so much that it caused them to lose contact with their rolling barrage. Research into the problem showed that German machine gunners had been noting the short limit of the artillery fire during the preparation and then moving into shell holes far enough into "No Man's Land" to avoid the incoming shells. These troops were ready and waiting with their machine guns when the infantry assault finally came. To catch these forward machine gunners, prior to the initial infantry assault of 1 November, the gunners supporting the 2nd Division spent the last ten minutes of the preparatory bombardment firing a heavy standing barrage that was drawn back 200 yards from the previous short limit,. During this barrage, the leading assault troops moved up as close to the standing barrage as possible, so that they could hit the first enemy positions within moments of the barrage's first lift forward.[19]

After the start of the infantry attack, the focus of nearly all the firepower in the corps was on covering the infantry attack. The rate of advance of the rolling barrage was tailored to match the type of terrain facing the infantry: 100 meters every four minutes on open ground, 100 meters every six minutes up steep slopes, and 100 meters every eight minutes through woods. Two batteries in each light battalion (75mm) were to shoot high-explosive shell, and the third was to superimpose shrapnel fire on the whole battalion sector, firing 200 meters ahead of the batteries shooting the high explosive. Other guns were dedicated to firing smoke shell throughout the barrage. The 155s were directed to fire "a series of standing barrages, lifting as the infantry advances so as to fall at least 500 meters beyond the rolling barrage." Following each infantry objective the rolling barrage was to halt, and the 155s were to shoot "heavy concentrations" on "all enemy organizations" within two kilometers of the new front line. Batteries of eight-inch howitzers were designated to further intensify this accompanying fire by adding another layer to the barrage, 500 meters ahead of the 155s, when the barrage rolled through certain strong points and wooded areas. The order made it clear that army and corps guns not assigned specific counterbattery missions were to continue supporting the infantry advance all the way through the third and final objective of the first day.[20]

To accomplish this massive fire plan, the 2nd Division received more artillery augmentation than in any previous attack. In addition to its own guns, it would have the support of two other divisional artillery brigades, the 1st FA Brigade of the 1st Division and the 67th FA Brigade of the 42nd Division (a total addition of four 75mm regiments and two 155mm regiments). The 2nd Division also received the dedicated support of two batteries of eight-inch howitzers and three batteries of 105mm guns, and would share with the 82nd Division to their right some seventeen batteries of 155mm howitzers and twenty-one batteries of 155mm guns from the corps and army artillery groups. All told, the infantrymen of the 2nd Division would be supported by over 300 guns and howitzers.[21]

Command of the three brigades of divisional artillery was given to the ranking artillery commander, General George G. Gatley from the 67th FA Brigade. However, more than with any other attack by the 2nd Division, the plans issued by the corps chief of artillery, in this case Brigadier General Dwight E. Aultman, had carefully scripted all artillery actions, even to the point of detailing exactly when each divisional artillery battalion would displace forward and to what position it was to move. Two hours after the infantry began their attack, the 2nd FA Brigade was to begin its movement forward by echelon, one battalion advancing at a time. The forward batteries were then to pick up the barrage, or otherwise support the infantry attack, after the first objective was reached. While exact forward positions were listed in the plan, the guiding principle was that "batteries were to be pushed as far forward as the tactical situation permits." The plan also directed that "previous to 'D' day," the 67th FA Brigade was to take up positions "as far advanced as compatible with conservation of the material and replenishment of ammunition," so that it would be sure to have the capability of firing the entire barrage without advancing. In fact, General Gatley procured enough of the long-distance 1917 Model "D" shell to enable both the 67th and 1st FA brigades to fire the barrage all the way through the final objective without having to advance their batteries.[22]

The divisional plan also had an additional measure of flexibility in its rolling barrage, which was inserted to limit the damage that would result from one of the more common problems in previous American attacks. The attack order stated that "should the infantry at any time be unable to keep up with the barrage, the barrage may be recalled to some well established line, to rest there until the infantry lines are adjusted."[23] This was easier to write into an attack order than to accomplish in the midst of an attack that was obviously not going well, but it does represent an important attempt to solve an all too common problem.

As at St. Mihiel, the corps instructions called for accompanying guns to be used in the attack. Aultman directed that "batteries or single guns will be detailed, after consultation with Division and Infantry Brigade Commanders, to accompany the advance of the infantry, for the purpose of firing on Machine Gun nests."[24] In this regard, the divisional plan specified that one battalion each from the 12th and 15th FA Regiments were "to be prepared to move forward for close support of the infantry as soon as the advance of the infantry warranted it." However, just as at St. Mihiel, the leadership of the 2nd Division had no intention of breaking up these battalions and assigning the pieces directly to individual infantry battalions. While still subject to requests for fire from any infantry unit, these battalions were to remain assigned to and under the control of their respective artillery battalion, regimental and brigade commanders.[25]

All told, it was the most comprehensive firepower employment plan that the 2nd Division, or any other AEF division, had ever seen. Everything was in place well in advance of the jump-off time, which was set for 5:30 AM, 1 November. During the artillery preparation, which began promptly at 3:30 AM, all the guns assigned to the corps opened up on their German targets, namely, batteries and those infantry positions that were able to deliver fire upon American infantry at their starting positions. Eighty gas projectors also fired lethal chemicals on known machine-gun nests, observation posts, and organized shell holes. Ten minutes before "H" hour, the standing barrage was put down in front of the German forward positions, as was a thick smoke screen to hide the American infantry moving up to the barrage. At 5:30 AM the barrage rolled forward, and the attacking infantry battalions of the 2nd Division began their advance (see Map 6.1), assisted by one company of fifteen tanks.[26]

Three regiments moved forward simultaneously, with the leading battalions of the 6th Marine, 5th Marine, and 23rd Infantry Regiments covering the four-kilometer-wide divisional front. The 9th Regiment advanced in support. Reports from across the front agree that the first German positions "did not offer a great deal of resistance." On the right, the 23rd Regiment "followed the barrage closely," cleared out the enemy

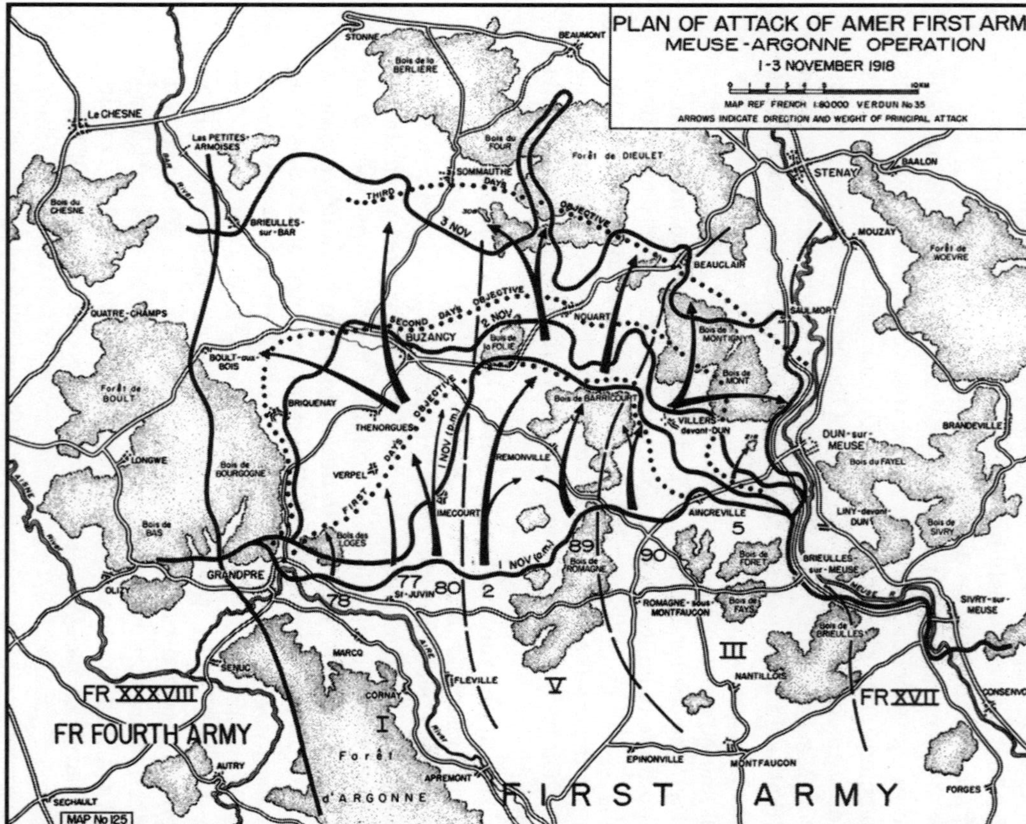

Map 6.1. Plan of attack and operations of the First American Army in the Meuse-Argonne, 1–3 November 1918. Notice the key central position and deep advances of the 2nd Division. Taken from *United States Army in the World War*, vol. 9, p. 373.

strong points in the Bois de Hazois, and reached the first objective at the scheduled time. At this point the divisional front shrank, and the 23rd Regiment moved to the left and took a support position behind the two Marine regiments. After the thirty-minute rest at the first objective, the Marines continued the advance, meeting only slight enemy resistance and taking "one objective after another on scheduled time." At the third and final objective the Marines consolidated their positions and sent patrols forward to the "exploitation line." The 9th and 23rd Regiments dug in behind them for the night.

The first day's attack was an unequivocal success; the division had broken through the enemy's main line of resistance, advanced nine kilometers, and had suffered only light casualties.[27] It was another well executed set-piece attack, in which all the limited objectives assigned were taken without any serious difficulties (see Map 6.2). Strategically, the division had achieved one of the main objectives of the campaign, as the German Army's critically important Metz-Sedan railroad line had been brought within range of heavy artillery fire. Speaking of the one-day advance of his division, Lejeune proudly claimed that it fundamentally changed the remainder of the offensive. For the Germans, the attack changed their defensive plan from one of "stubborn resistance, in a prepared position, to one of a series of holding engagements." For the Americans, the operation changed "from that of a prepared attack, with a defined objective, to one of exploitation of a success already gained."[28]

For the second time in less than a month, the artillery had proven itself more than equal to the tasks assigned it in the initial attack. The infantry was even more impressed with the artillery support of the 1 November assault than they were for the support of the first attack at Blanc Mont. The acting commander of the 3rd Brigade, Colonel Robert O. Van Horn, claimed that the preliminary preparation "was the most intense and best executed of any preceding any attack of the brigade [sic]."[29] General Neville, commanding the 4th Brigade, credited the "thorough" preparation with causing the withdrawal of a portion of the German forces.[30] The counterbattery effort also appears to have been successful, as some reports stated that the German artillery response was "weak" or "light," and others did not even deem it worthy of mention.[31]

As for the artillery fire accompanying the assault, a commander in the 3rd Brigade claimed that the "resistance of the enemy was shattered by the intensity and rapidity of the barrage fire" and that it was "most effective." Colonel Stone of the 23rd Infantry Regiment reported that "the initial attack was marked by a brilliant coordination of arms, the artillery laying down an absolutely smothering barrage which the infantry followed closely. . . . Attention must again be called to the wonderful work of our artillery in the first fighting. Our troops have never advanced behind a more perfect barrage."[32] These reports are confirmed by the

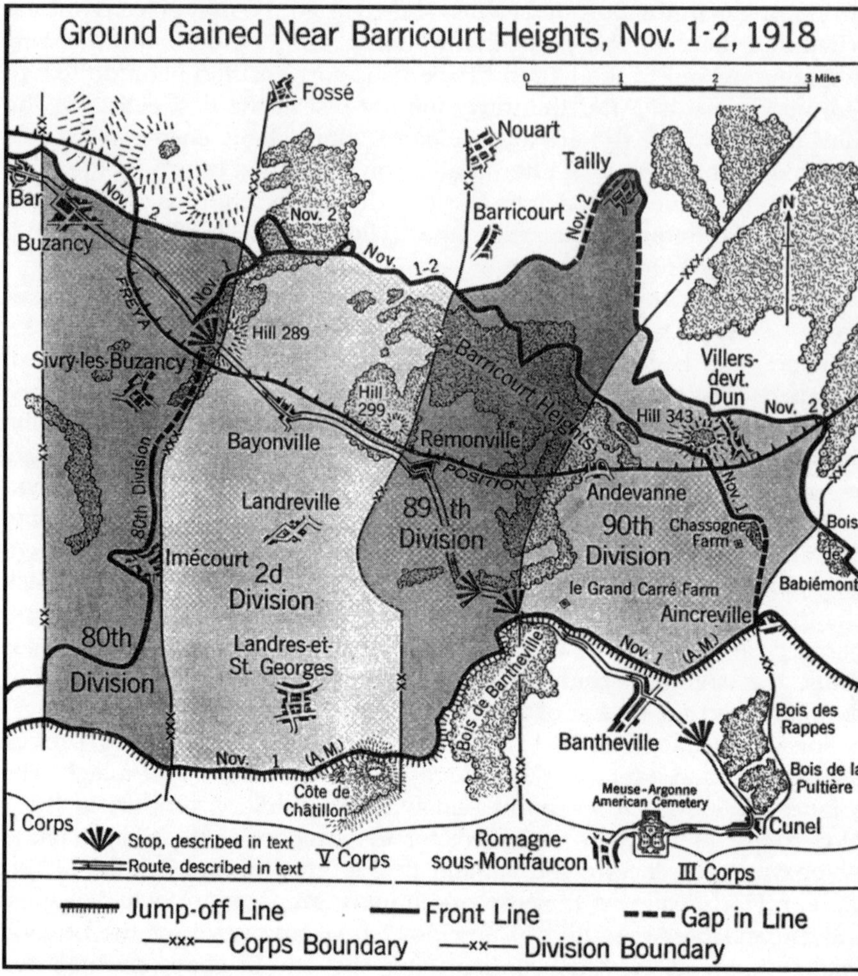

Map 6.2. Operations of the 2nd Division in the Meuse-Argonne on 1 November 1918. ABMC, *American Armies and Battlefields in Europe*, p. 276.

commander of the tank company that accompanied the initial assault. Captain C. H. Barnard reported that when his tanks advanced at "H" hour, they found that "the artillery work had been so thorough that there was very little wire to be cut and practically no machine gun resistance except for an isolated gun here and there which they destroyed."[33]

The 2nd FA Brigade also appears to have been very successful in moving forward during the initial attack. As early as 7:45 AM, General Bowley reported to the corps HQ that his first batteries were moving forward "as per the schedule." By 12:30 PM the advance battalions were in forward positions firing the rolling barrage and the other light battalions had been in motion for hours. The first battalion of 155s was on the road forward at 10:00 AM, and the other two were advancing before 5:00 PM. Every gun in the brigade was therefore in a forward position in time to guard the day's gains throughout the first night. While the 1st FA Brigade also moved forward that evening, the 2nd Division lost the support of the 67th FA Brigade, as that brigade was returned to its parent division.[34] This meant that by the beginning of the first night of the offensive, the division had already lost one third of its divisional artillery strength.

On the night of 1 November, V Corps ordered the 2nd Division to prepare to change directions and advance due west in order to make a flank attack on the German forces opposing the American division on its left. However, early the next morning the corps changed its mind and directed the division to continue advancing to the north. Due to these changes in orders, the division made no advance until that night.[35] After passing through the 4th Brigade, the 3rd Brigade formed its regiments in one long column of twos, with advance and flank guards, and made a bold advance under cover of darkness and rain some three kilometers to the "exploitation line" given in the first day's orders. The night march took the regiments through the Bois de Folie, which was still laced with German soldiers performing a rear-guard mission. By 6:00 AM, 3 November, the brigade was on its objective and was ready to continue the attack. Six hours later it had advanced another four kilometers, and found itself facing a German defensive position that was supported by artillery and Minenwerfers (trench mortars), at the south edge of the Bois de Belval.[36]

Although the supporting batteries from Colonel Davis's 15th FA had just moved to more advanced positions that day, the 3rd Brigade had no trouble arranging for fire to be brought on the German positions. Colonel Van Horn reported that the supporting guns from the 15th FA "shot up the resistance with good effectiveness."[37] After blasting the edge of the woods, the artillery fired a rolling barrage of sorts that advanced along a road through the woods, concentrating its fire to cover just 200 yards on each side of the road. That evening the 9th and 23rd Regiments again formed into columns of two and followed an advance guard through the woods, advancing some six kilometers that night. By midnight on 3 November, the leading elements of the division were some eight kilometers ahead of the divisions on their flanks. The success of the night march also proved the disorganization rampant in the forward German units at that time. Whole groups of German officers and soldiers were captured while sitting in well-lit farm houses or asleep in the woods, all totally unaware of the depth of the American advance.[38]

Finally, on the morning of 4 November, the Germans were able to piece together a strong line of resistance with well-organized machine-gun nests two kilometers south of the town of Beaumont. When Colonel

James C. Rhea, the new commander of the 3rd Brigade, realized his forces were up against a formidable enemy position, on difficult ground, with both flanks exposed, he reported the situation to General Lejeune, who ordered the brigade to continue the advance. Rhea claimed that he had just one battalion from the 15th FA in position to support the attack. While the attack succeeded in breaking the German line and in continuing the advance for about a kilometer, Rhea said it was only accomplished "at the expense of heavy casualties, the losses being in the neighborhood of 1,000 casualties, with a high percentage killed."[39]

Why was so little artillery support available for this attack? The division was in its fourth day of continuous operations, during which it had advanced more than twenty kilometers. It was straining its logistical sections to the limit, and the constant rain made that strain much worse. According to Rhea, the roads in his sector "were at first almost impassable, and after Nov. 3rd were entirely impassable." The artillery, and all of its ammunition, had to make its way forward either cross-country or by using the congested roads in other divisional sectors. A divisional inspector reported that all roads in the 2nd Division's sector were "axle deep in mud," and that he had personally witnessed a team of "22 horses hitched to one gun and they could just move it in the mud."[40]

A second factor contributing to the lack of artillery support was that on 4 November, the 4th Brigade moved out of its support role to advance into the divisional sector not covered by the 3rd Brigade's night marches. Therefore, the Marine brigade was using the batteries of Lieutenant Colonel John A. Holabird's 12th FA to support that movement.[41] The 1st FA Brigade had been returned to its parent division that day as well, which meant that while the leading infantry battalions for the 1 November attack had the dedicated support of more than three whole artillery brigades, just three days later the assaulting infantry units were reduced to scraping together as many guns as they could get from their one supporting regiment of light guns and half of the 17th FA.

To make matters worse, division inspectors had reported to the inspector general of the AEF, and through him to General Summerall, that the 2nd Division did not properly employ "accompanying guns" during the initial assault, as per the corps order. Pressured to rectify this irregularity, General Lejeune ordered that as of 3 November, accompanying artillery was to be detailed to the infantry, with two guns going to each battalion in the front line.[42] Therefore, those guns were probably not available to the regimental artillery commander at a time when he was called on to provide the greatest possible concentration of fire.

On 4 and 5 November both infantry brigades pressed towards the Meuse River, and the German forces opposing them fought rear-guard actions to enable the greatest number of German troops and materiel to get behind their new line of resistance across the river. From 6 to 10

November, the infantry brigades improved their positions on the west bank of the Meuse, all the while being subjected to heavy German artillery fire from across the river. The artillery brigade also established gun positions in preparation for supporting an eventual crossing of the river.[43]

Finally, late on 9 November, General Lejeune complied with corps directives and ordered his troops to cross the Meuse in two places the following night. General Neville was ordered to command a crossing force composed of troops from two V Corps divisions, the 2nd and 89th divisions. Neville requested that the orders be changed to allow for a daytime crossing at just one location, since he was convinced of the strength of the German positions on the east bank and because he had "little artillery with which to support even one crossing." The requests were denied.[44] The 12th FA and four batteries of the 17th FA were to support the northern crossing near Mouzon, which was thought of as the main effort. The 15th FA, along with the other two batteries of 155s, were assigned to the southern crossing near Letanne. The crossings were attempted at 9:30 PM, preceded by an hour-long artillery preparation. At Mouzon, the responding German fire was so strong that the engineers and infantrymen could not even get the temporary bridges across. At Letanne, the German gunners were slower in identifying the crossing point, and three battalions from the two attacking divisions were able to fight their way across the foot bridges and establish positions east of the river. However, the casualties sustained in these "surprise" attacks were heavy; General Summerall even called them "excessive."[45] Within hours of the crossing the armistice took effect, and the war was over. The 2nd FA Brigade had fired its last shell of the war.[46]

The attack of the 2nd Division in the third phase of the Meuse-Argonne campaign was viewed by many in the AEF as being the most successful American attack of the entire war. In eleven days it advanced some 29 kilometers, captured over 1,700 prisoners and 105 artillery pieces, and suffered 3,299 casualties.[47] However, just as with the previous offensives, an operational analysis demonstrates that the division, and the AEF, showed both its strengths and its weaknesses at different parts of the campaign.

There can be no doubt that the preliminary bombardment was effective, as noted above. The decision to pull back the barrage for the last ten minutes of the preparation demonstrated an ability to adapt to the changing tactics of the enemy, and may have been an important measure of the success of the initial assault.[48] Another factor that was crucial to the success of the attack was the use of massive quantities of gas in the preparatory bombardment to neutralize both German batteries and infantry strong points. Rexmond Cochrane, in his analysis of the American use of gas in the Meuse-Argonne, asserted that "the gas plan for the final assault [on Nov. 1] was the best that could have been devised, and it succeeded exactly as planned."[49]

Despite these achievements, probably the most significant innovation made in the final attack was the massive, overpowering rolling barrage, which incorporated both heavy and light artillery throughout its execution. While the 2nd Division and the AEF had previously used heavy guns to fire in concentrations in advance of the 75mm rolling barrage, the goal of that fire had always been to "soften up" known or suspected obstacles that the infantry was eventually going to meet. The attack on 1 November was the first to incorporate the heavy guns into a comprehensive infantry support fire plan that had as its goal the neutralization of "only those positions from which fire could be delivered upon our infantry at that moment." The use of this "thick" barrage was the result of a victory by those in the AEF who were convinced that, regardless of what General Pershing thought the infantryman could accomplish with just his rifle and bayonet, "the assault battalions must be covered by artillery and machine gun fire in all stages of the advance."[50] The results achieved by the 2nd Division on the first day attest to the effectiveness of that principle.

Covering the infantry attack with fire was much easier to arrange on the first day of the attack, and the 2nd Division proved the necessity for overwhelming fire power again in the later stages of the attack by demonstrating the costs of not being able to provide it. Just as at Blanc Mont, the number of guns assigned to the division became smaller and smaller as the attack progressed. When situations arose in which the attacking infantry needed strong, concentrated artillery support, as occurred with the 3rd Brigade on 4 November and the 4th Marine Brigade in its attempt to cross the Meuse, the divisional artillery brigade was often unable to provide adequate support on its own. This insufficient artillery support was not caused solely by a reduction in available firepower, but also by the compression in the amount of time allowed to properly prepare and coordinate the firepower that was assigned. For the Meuse crossings, senior officers in the division attempted to secure the use of allied aircraft to get updated photographs of the new German defenses so that the fire plan for the attack would adequately suppress enemy fire. Colonel Dan Moore, who replaced Bowley as commander of the artillery brigade, asserted that "if photos had been available for the preparation of the attack for the crossing of the Meuse, a much more intelligent bombardment of the enemy's machine gun and battery positions might have been undertaken, and the crossing probably would have been a success at Mouzon."[51] The lack of time and resources available to adequately support infantry attacks in the "exploitation" phase were factors that caused the 2nd Division trouble throughout the war.

The use of "accompanying guns" was one innovative attempt to provide artillery support to the infantry attacking in the "exploitation" phase. However, when General Aultman, the chief of artillery for V Corps, reported that the 2nd Division could only make "little use" of its accompanying guns and that "they were of no value to the artillery" during the period they were parcelled out to the infantry, it was an all too common description of the use of accompanying guns by American units.[52] Clearly, Lejeune and Bowley shared a dislike for accompanying guns and were convinced that despite the problems of attempting to employ masses of artillery after the initial assault, it was better to maintain a system that enabled the possibility of concentrating a "more effective volume of fire . . . on any point desired" than to break up the guns and hope that they would be used to provide some support at critical points in the attack.[53] Colonel Moore, Bowley's replacement, closed his operations report for the Meuse-Argonne campaign by asserting that attacking divisions get the "best results from artillery by placing artillery support in the hands of the artillery regiment and battalion commanders, who follow the advance closely and solve the problems of enemy resistance through direct liaison with the infantry."[54] However, keeping all the guns "close" and the liaison "direct" was easier said than done, and the division never fully met either challenge.

The infantry regiments of the 2nd Division made much better use of a different innovation to reduce their susceptibility to enemy firepower in the "exploitation" phase of the attack. The night marches were clearly an attempt by the attacking infantry to continue to make aggressive advances in the midst of the enemy, without having to measure their own firepower against that of the enemy. It was a brilliant and bold tactic, and each maneuver saved many lives and much hard fighting. However, such a tactic would have been suicidal just one month earlier had it been tried at Blanc Mont, or just about anywhere else on the Western Front. From 3 November on, the German forces in the Meuse-Argonne were attempting to withdraw to a position east of the river, and that is what allowed the night marches to be such a success.[55] It is important to realize that even though the German forces were in the midst of a rather disorganized withdrawal, they were able on two occasions to confront the advancing forces of the 2nd Division and inflict heavy casualties on them.

Nevertheless, the German forces that were hurled back some nine kilometers during the initial assault on 1 November were, while tired and depleted, neither disorganized nor seriously considering a withdrawal. Those forces were resoundingly defeated by a 2nd Division force, which proved, once again, that it could deliver an extremely effective set-piece attack. It was only in the later stages of the "exploitation" that the 2nd Division showed that it still was battling the troubles caused by a doctrine that was ill suited to its area of expertise. Even at the end of the war, Pershing's query from 8 August was still very pertinent to AEF operations in the "exploitation" phase of any attack—perhaps they were *still* "losing too many men," and perhaps the doctrine Pershing relied on deserved a share of the blame.

In its three final attacks, at St. Mihiel, Blanc Mont, and the Meuse-Argonne, the 2nd Division positively demonstrated what it had given strong hints of at Vaux and Soissons—that it could execute limited, set-piece attacks with great success. It succeeded against determined defenders who were in strong defensive positions, and did so without suffering excessive casualties. It even proved capable of successfully executing these attacks despite being given insufficient time to prepare (e.g., at Blanc Mont) and having the divisions on its flanks make unimpressive supporting advances (e.g., Blanc Mont and Meuse-Argonne). What also emerged, most notably in the final two offensives, was a desire on the part of the senior divisional officers to break from official AEF doctrine and fight for limited objectives by maximizing the use of artillery. On the negative side, the division struggled to make successful attacks immediately after the initial assault. Though the division may have been inclined to hold its new lines and make preparations for additional artillery-centered, set-piece attacks, the French and American high commands both compelled it to press ahead. These follow-on attacks nearly always led to excessive casualties, while yielding unimpressive gains.

NOTES

1. Major General Summerall, "Comments by the Corps Commander upon the Operations of the Fifth Army Corps," in V Corps Historical File, Entry 1118, RG 120, NA.

2. Trask, *Coalition Warmaking*, pp. 122–23, 149.

3. Allan R. Millett, *The General* (Westport, CT: Greenwood Press, 1975), p. 399.

4. The German Army used gas extensively in the 1918 spring offensives to neutralize both enemy infantry and artillery. In the Chemin des Dames offensive, 50 percent of the three million artillery rounds fired on the first day were gas rounds.

5. "The failure to use gas in the Argonne and on the 5th Corps front in the opening assault were therefore major blunders," according to Rexmond Cochrane, *The Use of Gas in the Meuse-Argonne Campaign, September–November 1918*, Study #10, U.S. Army Chemical Corps Historical Studies: Gas Warfare in World War I (Army Chemical Center, Maryland, 1958), p. 90. While Cochrane asserted that there was no sign of any shortage of gas shells for the First Army, Colonel Conrad Lanza, the chief of operations for the First Army Artillery during the campaign, claimed that plans to use great quantities of gas shell "had to be abandoned due to the fact that nothing like the amount of gas shell asked for could be had." The testimony of General Charles P. Summerall, a division and

corps commander in the campaign, supports Lanza's claim. He stated that "the Corps was at all times anxious to fire large quantities of non-persistent gas and the amount of gas so fired was only limited by the supply that could be obtained. See Lanza, *The Army Artillery, First Army*, p. 11, and Summerall, "Comments by the Corps Commander upon the Operations of the Fifth Army Corps," p. 4, in V Corps Historical File, Entry 1118, RG 120 NA.

6. Lanza, "Artillery Support," pp. 70–71.

7. "Report of Operations During the Period Nov. 1st to Nov. 11th," HQ 2nd FA Brigade, 26 December 1918, Folder 15, 2nd Division files, G-3 Reports, AEF/GHQ, Entry 270, RG 120, NA; Lieutenant William D. Bickham and Major William Burr, "Second Field Artillery Brigade History," p. 22, undated typescript, 2nd FA Brigade Historical File, Entry 1246-11.4, RG 120, NA.

8. The division received the first drafts of the First Army and V Corps plans on 24 October, Spaulding and Wright, *The Second Division*, p. 198; "Report of Operations, 3rd Brigade from October 17, 1918 to 2nd Nov, 1918," Folder 50, 2nd Division files, G-3 Reports, AEF/GHQ, Entry 270, RG 120, NA.

9. "Report of First Army," *USAWW*, 9: 327.

10. Major General Charles P. Summerall lecture, "Recent Operations from the Standpoint of Employment of Artillery with Particular Reference to Co-operation between the Infantry and the Artillery," p. 5, to the ACAS, on 16 December 1918, in Folder 14, ACAS Files, Entry 371, RG 120, NA.

11. Field Order No. 101, V Corps, 28 October 1918, *RSD*, vol. 1.

12. The farthest main line of resistance (south of the Meuse) for the Germans was the Freya Stellung, which was within six kilometers of the 2nd Division's jump-off position and as close as four kilometers to the 89th Division's front lines.

13. Cochrane, *The Use of Gas*, p. 4.

14. Lanza, "Artillery Support," p. 75; "Report of First Army," *USAWW*, 9: 327.

15. Major General Charles P. Summerall lecture, "Recent Operations from the Standpoint of Employment of Artillery with Particular Reference to Co-operation between the Infantry and the Artillery," pp. 1–2, to the ACAS, on 16 December 1918, in Folder 14, ACAS Files, Entry 371, RG 120, NA.

16. See Summerall lecture, "Recent Operations from the Standpoint of Employment of Artillery with Particular Reference to Co-operation between the Infantry and the Artillery," pp. 1–2, to the ACAS, on 16 December 1918, in Folder 14, ACAS Files, Entry 371, RG 120, NA.

17. Operations Order No. 5, "Plan of Attack," Artillery HQ, V Corps, 29 October 1918, V Corps Historical File, Entry 1118, RG 120, NA.

18. This number comes from General Lejeune. How many machine guns were involved throughout the whole of V Corps front is not known. *Reminiscences*, p. 384.

19. Spaulding and Wright, *The Second Division*, pp. 199–200.

20. Operations Order No. 5, "Plan of Attack," Artillery HQ, V Corps, 29 October 1918, V Corps Historical File, Entry 1118, RG 120, NA.

21. Summerall stated that the 2nd Division had 236 divisional guns and the 89th Division, 208 guns, and that they shared 152 corps guns, including the dedicated army groups that were put under corps direction. This was spread on an eight-kilometer front, giving the corps an artillery density of one 75mm gun

for each 29 meters, a 155mm or 8" howitzer for every 39 meters, and a 105mm or 155mm gun for every 69 meters of front. The division also had a battery (12) of 6" Newton Mortars, and a number of gas projectors. Memo, Summerall to Deputy Chief of Staff, AEF, 5 January 1919, V Corps Historical File, Entry 1118, RG 120, NA; Operations Order No. 5, "Plan of Attack," Artillery HQ, V Corps, 29 October 1918, V Corps Historical File, Entry 1118, RG 120, NA.

22. Operations Order No. 5, "Plan of Attack," Artillery HQ, V Corps, 29 October 1918, V Corps Historical File, Entry 1118, RG 120, NA; Spaulding and Wright, *The Second Division*, p. 200.

23. Field Order No. 49, HQ the 2nd Division, 0700, 31 October 1918, 2nd Division Historical File, Entry 1241-32.1, RG 120, NA.

24. Operations Order No. 5, "Plan of Attack," Artillery HQ, V Corps, 29 October 1918, V Corps Historical File, Entry 1118, RG 120, NA.

25. "Report of Operations, 2nd FA Brigade," Folder 15, 2nd Division files, G-3 Reports, AEF/GHQ, Entry 270, RG 120, NA.

26. Each battery in the first battalion of the 12th FA fired 400 rounds at enemy strong points and forward positions. "Report on advance of 1st Btn, 12 FA, Nov. 1," Folder 9, 2nd Division files, G-3 Reports, AEF/GHQ, Entry 270, RG 120, NA; Summerall lecture, "Recent Operations from the Standpoint of Employment of Artillery with Particular Reference to Co-operation between the Infantry and the Artillery," pp. 1–2, to the ACAS, on 16 December 1918, in Folder 14, ACAS Files, Entry 371, RG 120, NA; "Report of Operations from Sept. 26th to Nov. 8th, 1918," HQ V Corps-Artillery, 14 November 1918, V Corps Historical File, Entry 1118, RG 120, NA; Memo, "Projected Use of Gas Troops," 26 October 1918, 2nd Battalion, First Gas Regiment, to V Corps Commander, Folder 14, 2nd Division files, G-3 Reports, AEF/GHQ, Entry 270, RG 120, NA; Spaulding and Wright, *The Second Division*, p. 200.

27. General Neville, whose Marine battalions served as the assault troops for all three first-day objectives, estimated that his brigade suffered casualties of just 5 percent of the forces engaged on that day. Lejeune reported that "the prisoners taken exceeded the losses sustained on that day." "Operations Report, 4th Brigade, Marines, 24 Oct.–11 Nov.," Folder 87, 2nd Division files, G-3 Reports, AEF/GHQ, Entry 270, RG 120, NA; "Report of Operations, 3rd Brigade from Oct. 17, 1918 to 2 Nov. 1918," Folder 50, 2nd Division files, G-3 Reports, AEF/GHQ, Entry 270, RG 120, NA; "Operations Report of the 2d Division," *RSD*, vol. 1.

28. "Operations Report of the 2d Division," *RSD*, vol. 1.

29. "Report of Operations, 3rd Brigade from Oct. 17, 1918 to 2 Nov. 1918," Folder 50, 2nd Division files, G-3 Reports, AEF/GHQ, Entry 270, RG 120, NA.

30. "Operations Report, 4th Brigade, Marines, 24 Oct.–11 Nov.," Folder 87, 2nd Division files, G-3 Reports, AEF/GHQ, Entry 270, RG 120, NA.

31. No unit in the 3rd Brigade even mentioned German artillery fire during this attack. "Records of Telephone Messages Sent and Received, Nov. 1st to Nov. 11th, inclusive," V Corps Historical File, Entry 1118, RG 120, NA; "Operations Report, 4th Brigade, Marines, 24 Oct.–11 Nov," Folder 87, 2nd Division files, G-3 Reports, AEF/GHQ, Entry 270, RG 120, NA.

32. "Report of Operations," 3rd Brigade, Nov. 2–11 and "Report of Operations, Nov. 1–11, 23d Infantry," 27 November 1918, found in "Report of Units of 2nd

Division, Meuse-Argonne Offensive, No. 1–11, 1918," Folder 15, 2nd Division files, G-3 Reports, AEF/GHQ, Entry 270, RG 120, NA.

33. Of the fifteen tanks that started with the 2nd Division, only six reached even the first objective. The rest either broke down or were "hung up" in tank traps, ditches, or deep shell holes. The 4th Brigade said they were "ably supported by the tanks" while the 23rd Infantry claimed that the "tanks failed to keep up and were of little assistance." See "Report of Operations, 1st Provisional Tank Company, 1st Brigade Tanks," 1 November 1918, in 2nd Division Historical Files, Entry 1241-33.0, RG 120, NA; "Operations Report, 4th Brigade, Marines, 24 Oct.–11 Nov," Folder 87, 2nd Division files, G-3 Reports, AEF/GHQ, Entry 270, RG 120, NA; "Report of Operations, Nov. 1–11, 23d Infantry," 27 November 1918, found in "Report of Units of 2nd Division, Meuse-Argonne Offensive, No. 1–11, 1918," Folder 15, 2nd Division files, G-3 Reports, AEF/GHQ, Entry 270, RG 120, NA.

34. "Records of Telephone Messages Sent and Received, Nov. 1st to Nov. 11th, inclusive," V Corps Historical File, Entry 1118, RG 120, NA; "Report of Operations during the Period Nov. 1st to Nov. 11th," HQ 2d FA Brigade, Dec. 26, 1918, Folder 15, 2nd Division files, G-3 Reports, AEF/GHQ, Entry 270, RG 120, NA.

35. "Operations Report," 2d Division, *RSD*, vol. 1.

36. "Report of Operations," 3rd Brigade, Nov. 2–11, in "Report of Units of 2nd Division, Meuse-Argonne Offensive, No. 1–11, 1918," Folder 15, 2nd Division files, G-3 Reports, AEF/GHQ, Entry 270, RG 120, NA; Spaulding and Wright, *The Second Division*, pp. 208, 210.

37. One battalion from the 17th FA had been assigned to the 15th FA for advanced operations on this day and took part in the shelling of the German positions in the woods. "Report of Operations, Nov. 1–11, 9th Infantry Regiment," 29 November 1918, Folder 15, 2nd Division files, G-3 Reports, AEF/GHQ, Entry 270, RG 120, NA; Spaulding and Wright, *The Second Division*, pp. 208–9.

38. "Report of Operations," 3rd Brigade, Nov. 2–11, in "Report of Units of 2nd Division, Meuse-Argonne Offensive, No. 1–11, 1918," Folder 15, 2nd Division files, G-3 Reports, AEF/GHQ, Entry 270, RG 120, NA; "Operations Report," 2d Division, *RSD*, vol. 1.

39. "Report of Operations," 3rd Brigade, Nov. 2–11, in "Report of Units of 2nd Division, Meuse-Argonne Offensive, No. 1–11, 1918," Folder 15, 2nd Division files, G-3 Reports, AEF/GHQ, Entry 270, RG 120, NA.

40. "Report of Operations," 3rd Brigade, Nov. 2–11, in "Report of Units of 2nd Division, Meuse-Argonne Offensive, No. 1–11, 1918," Folder 15, 2nd Division files, G-3 Reports, AEF/GHQ, Entry 270, RG 120, NA; Report of 14 November 1918, in Inspection Reports of Armies, Divisions and Corps, Entry 590, RG 120, NA.

41. "Operations Report, 4th Brigade, Marines, 24 Oct.–11 Nov," Folder 87, 2nd Division files, G-3 Reports, AEF/GHQ, Entry 270, RG 120, NA.

42. "Report of Operations from Sept. 26th to Nov. 8th, 1918," HQ V Corps-Artillery, 14 November, 1918, V Corps Historical File, Entry 1118, RG 120, NA; Report of Colonel J. C. Johnson, Inspector General, "Operations and Conditions, 2d and 89th Divisions, V Corps, 1–3 Nov. 1918," and Memo, Major General A. W. Brewster, AEF Inspector General to V Corps Commander, subject: "Use of Artillery 2nd Division," both in Artillery folder, Entry 590, RG 120, NA.

43. "Report of Operations during the Period Nov. 1st to Nov. 11th," HQ 2d FA Brigade, Dec. 26, 1918, Folder 15, 2nd Division files, G-3 Reports, AEF/GHQ, Entry 270, RG 120, NA; "Report of Operations," 3rd Brigade, Nov. 2–11, in "Report of Units of 2nd Division, Meuse-Argonne Offensive, No. 1–11, 1918," Folder 15, 2nd Division files, G-3 Reports, AEF/GHQ, Entry 270, RG 120, NA; "Operations Report, 4th Brigade, Marines, 24 Oct.–11 Nov," Folder 87, 2nd Division files, G-3 Reports, AEF/GHQ, Entry 270, RG 120, NA.

44. "Operations Report," 2d Division, *RSD*, vol. 1; "Operations Report, 4th Brigade, Marines, 24 Oct.–11 Nov," Folder 87, 2nd Division files, G-3 Reports, AEF/GHQ, Entry 270, RG 120, NA; Spaulding and Wright, *The Second Division*, pp. 219–20.

45. Memo, Summerall to Deputy Chief of Staff, AEF, 5 January 1919, p. 6, V Corps Historical File, Entry 1118, RG 120, NA.

46. "Operations Report, 4th Brigade, Marines, 24 Oct.–11 Nov," Folder 87, 2nd Division files, G-3 Reports, AEF/GHQ, Entry 270, RG 120, NA; Spaulding and Wright, *The Second Division*, pp. 219–20.

47. Again, the ABMC total for casualties is slightly lower, 3,282; the figure in the text is taken from the report of 2nd Division Commander (Major General John A. Lejeune) to Assistant Chief of Staff, G-3, GHQ, AEF, Subject: "Operations Information," 30 December 1918, in Reports and Other Records Regarding the History of AEF Units, Divisional Skeleton Histories, Entry 443, RG 120, NA.

48. The U.S. 80th Division, on the left flank of the 2nd Division on 1 November, was stopped immediately by machine guns that had slipped beyond the artillery preparation and rolling barrage. The division inspector who investigated the poor first-day showing of that division reported that they should have done just what their neighbors to the right did. Memo for Inspector General, AEF, entitled "Observations with 1st and 5th Corps," 8 November 1918, in "Inspection Reports of Armies, Divisions, and Corps," Entry 590, RG 120, NA.

49. Cochrane, *The Use of Gas*, p. 91.

50. Summerall, "Recent Operations from the Standpoint of Employment of Artillery with Particular Reference to Co-operation between the Infantry and the Artillery," pp. 2–4, lecture to the ACAS, on 16 December 1918, in Folder 14, ACAS Files, Entry 371, RG 120, NA.

51. Colonel Moore replaced General Bowley on 3 November, as the latter was promoted to be the chief of artillery for VI Corps; "Report of Operations during the Period Nov. 1st to Nov. 11th," HQ 2d FA Brigade, Dec. 26, 1918, Folder 15, 2nd Division files, G-3 Reports, AEF/GHQ, Entry 270, RG 120, NA.

52. General Lejeune said much the same thing when he asserted that "no material advantage was gained" by his employment of accompanying guns. He was convinced that "more satisfactory results are obtained when all of the artillery remains under the control of the Artillery Commander and the 37mm and Stokes mortars are used to best advantage by the infantry." Memo for Inspector General (IG), AEF, 8 November 1918, subject: "Observations with First and Fifth Corps with respect to Artillery in recent operations," Entry 590, RG 120, NA; "Report of Operations from Sept. 26th to Nov. 8th, 1918," HQ V Corps-Artillery, 14 November, 1918, V Corps Historical File, Entry 1118, RG 120, NA.

53. General Bowley told an IG official that he "did not believe in this use of accompanying guns." Report of Colonel J. C. Johnson, Inspector General, "Op-

erations and Conditions, 2d and 89th Divisions, V Corps, 1–3 Nov. 1918," and Memo, Major General A. W. Brewster, AEF Inspector General, to V Corps Commander, subject: "Use of Artillery 2nd Division," both in Artillery folder, Entry 590, RG 120, NA.

54. "Report of Operations during the Period Nov. 1st to Nov. 11th," HQ 2d FA Brigade, Dec. 26, 1918, Folder 15, 2nd Division files, G-3 Reports, AEF/GHQ, Entry 270, RG 120, NA.

55. Spaulding and Wright, *The Second Division*, p. 210.

Conclusion: Between Pershing and the Revisionists—Artillery, the 2nd Division, and the AEF Style of Warfare

> The artillery was considered an auxiliary, sometimes useful, never necessary, and sometimes a nuisance. . . . There was an overconfidence as to the possibilities of infantry advancing, which was not born out by experience. . . . We tried to bring about Open Warfare; so did every other nation strong enough to adopt the offensive. We never succeeded, and no other nation succeeded. . . . [1]
>
> Colonel Conrad H. Lanza

The 2nd Division ended the war in the same condition that it finished its first operations—battered, but victorious. These two characteristics say a number of things about the operational capabilities of the division itself and of the AEF as a whole. The combat operations of the 2nd Division demonstrated both the strengths and the weaknesses of artillery employment in the division, and by extension in the AEF. In its earliest operations, the 2nd Division learned that there was a definite connection between the effectiveness of the artillery support provided and the success of an infantry attack. This proved to be more or less true, not just for the 2nd Division, but for the rest of the AEF as well.

This book has investigated artillery employment in the 2nd Division as a means of exploring the combat effectiveness of the AEF as a whole. In this respect it is necessary to determine where this analysis fits within the historiography of AEF operations, as well as with some initial assessments made by AEF officers immediately following the war. It will also be useful to note the impact of the French and British on the AEF, and ultimately to see whether the 2nd Division reflected a particular style of warfare that characterized the fighting of the AEF. If so, where does this style fit in with the historical warmaking strategies in American military tradition?

ARTILLERY EMPLOYMENT IN THE 2ND DIVISION AND THE AEF

Artillery employment has received remarkably little attention in the existing scholarship on the AEF. Yet, the effective, coordinated use of artillery was a major contributor to the operational effectiveness of any World War I force. The casualty rates nearly prove this single-handedly. The Medical Department of the U.S. Army determined that over 87 percent of all AEF combat casualties were caused by artillery fire of one form or another.[2] While it might have been the enemy machine guns that pinned attacking infantry to the ground, it was the artillery fire that quickly zeroed in on the attackers and did most of the killing and wounding. It was the job of friendly artillery to neutralize enemy fire. This was especially true when the attacking forces were not heavily supported by tanks, which was the case in most AEF attacks, so that the supporting artillery had to ensure that both the enemy machine guns and artillery were prevented from accurately firing.

This analysis of the 2nd Division's use of its 2nd FA Brigade, and of the artillery regiments temporarily assigned to it, demonstrates that artillery employment in the AEF was in many ways a mixed bag. After the division learned the importance of covering its attacking infantry with overwhelming artillery fire, a lesson it learned at great cost near Chateau-Thierry and Soissons, it demonstrated that it possessed the planning and technical skills necessary to deliver excellent artillery support for the initial assault of every offensive. The operations reports from the divisional infantry commanders, as well as the actual results attained, prove this to be the case. Surprisingly, the 2nd Division demonstrated that it had mastered these skills on a modest scale as early as the 1 July 1918 attack on Vaux, after having been in active combat for less than one month.

However, just weeks after the "mechanically perfect" attack on Vaux, the division showed at Soissons that it was not capable of providing adequate artillery support in a mobile, "open-warfare" environment, after the initial assault. Though some improvement was noted during the later stages of the Blanc Mont and Meuse-Argonne attacks, the 2nd Division continued to struggle in its efforts to effectively employ sufficient artillery support after making a successful initial assault. Invariably, as the division continued its attacks during a given offensive, it proved increasingly unable to coordinate and deliver the kind of overwhelming artillery support that had enabled it to make its impressive initial gains with slight losses.

As for the rest of the AEF, it appears that this pattern of initial success and subsequent difficulty extended to many other divisions. In the first phase of the Meuse-Argonne attack, for example, the American First Army lurched forward some ten kilometers in the first few days, but then came to a halt as the infantry attempted to advance beyond the zone covered by the initial artillery fire plan. An American artillery brigade commander confirmed that follow-on artillery support was a problem in other AEF divisions by asserting, "our artillery was fine in carrying out an operation which could all be planned and laid out beforehand, but that *when it was required to act in situations which had not been foreseen and planned beforehand, it fell down*"(emphasis added).[3] This decrease in effective artillery support resulted in the higher casualty figures and the more modest, or even insignificant, advances that characterized nearly all subsequent infantry attacks.

THE HISTORIOGRAPHY OF AEF COMBAT EFFECTIVENESS

This description of operational effectiveness does not square with either of the two main schools of thought that are prevalent in American World War I historiography. The first school has been dominated by the influence of General Pershing, most notably through his "Final Report," memoirs, and service as chairman of the American Battle Monuments Commission.[4] Pershing's version of AEF operational effectiveness is best presented in his own words. He described the AEF artillery forces this way:

> Rushed to France with but the most rudimentary training and here assigned new materiel and unaccustomed methods, *they overcame all difficulties by their energy, determination and devotion to duty*, affording to the Infantry that powerful support without which success would have been impossible. (emphasis added)[5]

As for the effectiveness of the AEF as a whole, Pershing came very close at times to asserting that the AEF was the best army in Europe. During the war he asserted to the War Department that "our troops are the best in Europe and our staffs are the equals of any."[6] After the war, Pershing tended to downplay or plainly ignore any operational shortcomings, and portrayed his command as "a powerful and smooth-running machine." He claimed that as early as March 1918, the AEF had combat divisions that "were equal to any demands of battle action," and that during the Aisne-Marne counteroffensive, "the tactical handling of our troops under these trying conditions was excellent throughout the action."[7] Of course, these were all very generous, if not quite accurate, assessments.

The picture painted by General Pershing gives the impression that the AEF executed all of its combat operations with consummate skill and efficiency, and that this battlefield expertise (particularly in the Meuse-Argonne campaign) was the primary factor in bringing the war to a close before 1919.[8] A detailed analysis of the operations of the 2nd Division, which was among the best of the AEF divisions, does not bear out Pershing's assertions. Pershing made no mention of the clear pattern that came to characterize so many American attacks, in which initially successful attacks were consistently continued well past the point at which they became excessively costly. Possibly Pershing was reticent to address this issue because he knew that the blame for the costly subsequent attacks would make its way to a commander-in-chief who had insisted on a doctrine based primarily on the thoroughly discredited concept of "self-reliant infantry."

Pershing's version of AEF operational effectiveness, which focused on battlefield successes, the ground gained, the challenges faced and overcome, and the ultimate victory, cast a long shadow on the American historiography of the AEF. As late as the 1960s, American military historians were still using Pershing's account as the basis of their works. Edward M. Coffman's *The War to End All Wars*, Harvey A. DeWeerd's *President Wilson Fights His War*, as well as other scholarly works, may have added a measure of criticism to various aspects of Pershing's benevolent account, but by and large they accepted the basic tenets of his version. However, the honeymoon was not destined to last forever.

Starting in the 1980s, a wave of revisionist works began to surface that focused on the American war effort and the operational effectiveness of the AEF. Historians such as James W. Rainey, Donald Smythe, Paul F. Braim, and David F. Trask told a very different version of events. Both in content and in tone, these writers described the AEF in rather unfavorable terms. In an article in *Parameters*, the journal of the U.S. Army War College, Rainey claimed that the AEF was "Partially schooled in trench warfare and in an ill-defined doctrine of open warfare, but well trained in neither." Therefore, the American forces were "unable to perform in concert as a well-oiled machine." He summarized AEF combat effectiveness by asserting that, "In having to grope its way to victory, the AEF succeeded not because of imaginative operations and tactics nor because of qualitative superiority in open warfare, but rather *by smothering German machine guns with American flesh*" (emphasis added).[9]

In his fine analysis of the Meuse-Argonne campaign, *The Test of Battle*, Braim asserted that "judged in its ability to shoot, move, and communicate . . . the performance of the First Army . . . must be given a fairly low rating." As a detailed examination of a single offensive, Braim's book tends to discount or ignore American involvement in all other major operations, and gives the impression that the "low rating" applied to the entire AEF.[10] Trask's work, *The AEF and Coalition Warmaking, 1917–1918*, is the latest, and in many ways the most critical, of both Pershing and the AEF. Trask stated that the AEF "could not fight efficiently on a large scale," that the First Army's performance was "unimpressive," and that it "suffered from incomplete organization, inadequate training, unsound logistical support, and inexperienced commanders and staffs." The Meuse-Argonne campaign was described as a "dubious victory," based on nothing more than "massed infantry attacks without effective coordination with other arms."[11]

The revisionists rely heavily on the AEF's high divisional casualty rates to support their claim that the AEF was not an effective fighting force.[12] It is true that AEF divisions often suffered very heavy casualties in their attacks, and it is therefore a valid point. However, the revisionists rarely provide the details necessary to identify the many external factors that often contributed to those high casualty rates. For example, there is usually no mention of the negative external influences of the French Army when American divisions fought under its control, and side by side with French divisions. Many of these negative influences led either directly or indirectly to higher casualties. The revisionists have tended to ignore these external factors, while regularly pointing to the heavy casualty rates as prima facie evidence of AEF combat inefficiency. On this issue, as with others, the reality is much more complicated than the revisionists have led readers to believe.

A NEW PERSPECTIVE

The revisionist account is not so much inaccurate as incomplete. There is no doubt that the revisionists were justified in exploding the myth that Pershing's army was a "smooth-running machine." They finally uncovered the sanguinary truth that, though the AEF was a part of the Allied forces that won the war, it finished the war far more bloodied than necessary. In many cases American divisions suffered heavy casualties that were caused primarily, if not exclusively, by internal problems. The fact that the war ended just six months after the AEF began to fight, along with the relatively low final casualty figures in comparison to those of the other armies on the Western Front, contributed to the illusion that the AEF's battlefield expertise was the deciding factor in the war.[13] The revisionists have helped correct this fallacy. Nevertheless, the revisionist portrayal of the AEF as a basically incompetent military force leaves no room for the considerable operational capabilities and undeniable achievements of its best divisions, such as the 2nd Division.

This book has described in detail each one of the five major operations in which the 2nd Division was engaged in the war. The analysis ultimately presents the picture of an AEF division that does not mesh with the awesome military force described by Pershing and those of his school. Nor does this analysis support the revisionist assertion that the AEF was a thoroughly incompetent and inefficient battle force. Rather, it suggests that the reality concerning the military effectiveness of the AEF lies between the two predominant schools.

The 2nd Division showed its inexperience near Chateau-Thierry as it struggled to drive the German forces out of Belleau Wood. There the division learned how important it was to provide its attacking infantry with overwhelming artillery support. It proved that it was capable of providing that support in its final attack on Belleau Wood, as well as in the attack of Vaux. At Soissons, the division demonstrated that though it could not provide the kind of flexible artillery support required to accomplish mobile operations without absorbing heavy casualties, it was still capable of outpacing some of the best troops in the French Army. Even in its earliest operations, the 2nd Division demonstrated that though it was inexperienced, it possessed a measure of technical skill and military competence. It had weaknesses, but it also had strengths.

These truths were confirmed in the division's final three operations at St. Mihiel, Blanc Mont, and in the Meuse-Argonne. In each attack the division proved it was capable of driving through well-fortified enemy lines, holding the new exposed positions against strong counterattacks, and doing so without suffering heavy casualties. It was exceptional assaults like those made by the 2nd Division at Blanc Mont and in the Meuse-Argonne that led General Summerall to boldly proclaim that "any position, however strong, can be neutralized and captured, with sufficiently powerful fire."[14] The revisionists have tended to obscure or ignore the fact that some units in the AEF had reached this level of proficiency after a relatively short period of combat. It was only when additional attacks were ordered that the 2nd Division displayed an inability to successfully advance without absorbing excessive casualties.

When those subsequent attacks were made, they usually resulted in smaller gains and higher casualties. This was true for the 2nd Division, as well as other divisions in the AEF, and even for divisions in the other Allied armies. It was probably this truth that caused General Summerall to assert that, "invariably the lack of sufficient superiority of fire is paid for in losses to the infantry and even in failure."[15] The truth of Summerall's statement was demonstrated in the later phases of the 2nd Division attacks near Soissons, St. Mihiel, Blanc Mont, and the Meuse-Argonne. As this work has chronicled, in each of those operations the 2nd Division struggled to one degree or another to provide the "sufficient superiority of fire" after the successful initial assault phase had ended.

What emerges is a pattern that accurately characterizes AEF offensive capability. There was an initial period of effectiveness, in fact a relatively high level of effectiveness in many cases, which was followed by a period of decreasing efficiency. Sometimes quickly, as at Soissons, and sometimes slowly, as in the Meuse-Argonne, but always eventually, the division reached a point of offensive ineffectiveness. Unfortunately, the two historiographical groups analyzing AEF combat experiences have focused primarily on just one phase of this pattern. On the one hand, General Pershing, and those who accepted his account, focused almost exclusively on the initial periods of success, while on the other hand, the revisionists have zeroed in on the latter periods of failure. However, as this book has demonstrated in the detailed analysis of the 2nd Division, the most complete picture of AEF operational effectiveness requires that all parts of the AEF offensive pattern be observed and analyzed.

THE IMPACT OF THE ALLIED ARMIES ON THE 2ND DIVISION AND THE AEF

Any attempt to understand the American combat experience in World War I must also explore the considerable influence exerted by the French and British forces, which played such a large role in training the AEF and, in many instances, commanding it in battle. Concerning Allies, this analysis has focused primarily on the impact of the French Army, since all AEF artillery training was accomplished under French tutelage, and since the 2nd Division was under the control of French army and corps commanders in three of its five major operations. However, some American divisions did the bulk of their infantry training with the BEF and subsequently fought under British control.[16] Despite Pershing's pride in refusing full amalgamation of American soldiers into the Allied armies, the French and British Armies exerted great control over the training, and even the operational activity, of nearly every American division at one time or another.

In fact, without the technical training received from the French and British Armies, no AEF division would have achieved the admirable level of efficiency that some did attain. This is especially true concerning combat in the trench warfare environment. As was the case for the 2nd Division, it was the French instructors who first stressed the absolute necessity of using overwhelming artillery support for limited attacks, and it was those same instructors who taught the division the technical skills required to carry out such operations. Pershing feared that having his forces trained by the French Army would ruin their aggressive spirit and prepare them for the wrong type of fighting. In the end, not only did his fears appear to be unfounded, but the French Army taught the American divisions those skills best suited to the attritional warfare in which they were eventually forced to engage.

However, operationally, the impact of the French Army was less beneficial. In fact, in many cases, the decisions of French commanders and the actions of French units had direct, negative consequences on American operations. For example, the 2nd Division was often hurriedly thrown into the front line by the French Army, ordered to attack without sufficient time to make proper preparations, given inadequate information on the battlefield, and provided with insufficient artillery augmentation. Additionally, in two of the joint attacks (Soissons and Blanc Mont) the 2nd Division suffered increased casualties because it attacked faster and further than the French divisions on both of its flanks.

These negative external factors cannot be ignored in assessing American divisional combat effectiveness. The fact that a relatively inexperienced division like the 2nd Division was able to advance as fast and as far as it did at Soissons and Blanc Mont, despite the negative influences of the French Army, lends further support to the position advanced in this work, that the AEF was, at least in parts, a more capable force than many modern scholars have described.

THE NEW PERSPECTIVE AND INITIAL POSTWAR IMPRESSIONS

Although the perspective of this book has been lacking in most American postwar historiography, it was actually expressed by some individuals directly after the war. Several officers in the AEF experienced and recognized this offensive pattern of initial success and subsequent difficulty. Colonel Lanza noted that when the infantry was supported by an overwhelming superiority of fire throughout the attack, it could advance without suffering excessive casualties, but that subsequent attacks, made without sufficient support, led to severe casualties and insignificant advances. He claimed that "given sufficient guns and ammunition, we can neutralize hostile artillery and machine guns, for the time needed to assault a designated position."[17] However, as for the subsequent follow-on attacks, Lanza criticized senior AEF officers for "relying on out of date training regulations which had emphasized the idea that the infantry was an independent arm and dependent on no other arm or service." Lanza wrote that this led "to the mistaken belief" that the infantry "could always advance no matter whether or not it had [heavy artillery] support." He accurately pointed out that this "overconfidence as to the possibilities of infantry advancing . . . was not born out by experience."[18]

Of all AEF senior officers, General Summerall might have had the clearest understanding of the offensive pattern and how it could be improved. Summerall made numerous attempts to explain this recurring offensive sequence to others in the AEF and in the U.S. Army. His reports on the operations of V Corps in the Meuse-Argonne, and his postwar lectures to other officers, both describe the issue in great detail. Accordingly, he advocated a more set-piece, artillery-centered approach for future offensives, an approach that assigned "successive objectives" as a means of taking the "successive lines of resistance." In this way, Summerall hoped to turn future American offensives into a repetitious series of those successful initial assaults.[19]

"THE AMERICAN WAY OF WAR" AND THE AEF STYLE OF WARFARE

Pershing did not appear to correctly identify this pattern, though he was surely aware of the high casualty figures that resulted from the later phases of the attacks. To try to answer why he did not consider making adjustments to this attacking pattern, it might be helpful to employ an argument offered by the American military historian Russell F. Weigley. In *The American Way of War*, Weigley applied conceptions of fundamental military strategy as explained by the military historians and theorists Carl von Clausewitz and Hans Delbrück, and asserted that America has used two distinct national military strategies throughout the prosecution of its wars—the strategy of attrition and the strategy of annihilation. The strategy of attrition, employed when America was "weak," such as during the War of Independence and the War of 1812, was an attempt to secure limited objectives with limited victories. From that point onward, as its military might grew, America developed a penchant for the second, more challenging strategy, that of annihilation. This meant that America sought the destruction of the enemy's military power, as well as the complete overthrow of the existing enemy government.[20]

While the strategy of annihilation was initially demonstrated in the early Indian wars, it was put to its first great test in the American Civil War. Weigley states that "the Civil War tended to fix the American image of war from the 1860s into America's rise to world power at the turn of the century, and it also suggested that the complete overthrow of the enemy, the destruction of his military power, is the object of war." Eventually, Weigley claims, "the strategy of annihilation became characteristically the American way in war."[21]

The significance of this reigning military theory for the eventual attitudes and doctrines of the AEF should not be missed. Pershing was born the year before the Civil War began, and his military education and training in the American Regular Army was heavily influenced by the strategy of annihilation that had solidified itself in American military tradition during the second half of the nineteenth century. Most other officers in the American Army were equally steeped in this tradition. Consequently, when the AEF arrived in France, it was commanded by men who were intent on achieving the decisive military victory that they believed was part and parcel of the strategy of annihilation. Pershing's hatred of the limited, attritional, trench warfare that characterized the Western Front in 1917 has already been noted. It seems highly probable that this sentiment, as well as his insistence on continuing to attempt to gain a decisive victory over the German Army, was based on the reigning strategic theory of the past half century. Pershing was intent on waging a war of annihilation, and his offensive theories were based on that premise.

A second theme of Weigley's work is that "technological and social developments" in the modern era have made it extremely difficult for even the most powerful nations to achieve the kind of decisive victory required by the strategy of annihilation. As this has been true in the arena of national military strategy, so it was true at the operational level of combat on the Western Front during World War I. After the lines stabilized in late 1914, no nation possessed the military power required to wage an offensive campaign that was decisive enough to bring a strategy of annihilation to fruition. Nevertheless, despite years of failure on the part of the massive European armies on the Western Front, General Pershing, and no doubt many other American officers, believed that it could be, and had to be done.

Pershing surely thought that the United States, with its large population and massive industrial potential, could field an army capable of waging the war of annihilation called for by American military tradition. Whether the United States of 1917 might have been able to field such a force after a lengthy period of mobilization and training will never be known, as it did not get the chance. What we do know is that the AEF did not become such an army—nor for that matter did any of the other armies of World War I. Unfortunately, Pershing seemed unwilling or unable to admit that to be the case. As a result, the AEF was forced against the will of its senior commanders and staff officers into waging a war of attrition, and ended up carrying out a much less effective war of attrition than it could have waged.

This book has demonstrated that the AEF, though not capable of achieving the decisive results Pershing demanded, did possess the enthusiasm, courage, and technical abilities necessary to carry out a very effective series of limited, set-piece attacks. Of course, a campaign based on a series of limited, set-piece attacks was a strategy of attrition, and that was anathema to most senior American officers like Pershing, and in direct opposition to the traditional American way of war. The strategy of attrition was one driven by weakness and, in Pershing's mind, a lack of aggressiveness as well. Pershing was convinced that his nation and its military force would be both strong and aggressive.

However, the AEF was in fact "strategically" weak, as were all the other armies on the Western Front, in those areas required to effectively carry out a decisive, annihilating offensive: communications, logistics, mobility, flexibility, force training, and the ability to combine speed with firepower. Even the German Army, which in the spring of 1918 came closer to achieving a decisive battlefield victory than any other force during the war, was too weak in those critical areas to effectively carry out a decisive campaign. And the German Army suffered very severe losses in its final attempts to do so.

One contributing cause of the weakness that worked against Pershing's plan of building an operationally decisive force was the ambivalent nature of the instruction given to even the best trained AEF divisions. Pershing demanded that his troops and divisions be trained in a manner that he was convinced would best prepare them to carry out aggressive, decisive operations. Whether Pershing's theories on this training, such as his focus on "open warfare" and "self-reliant infantry," were capable of giving rise to such operational capability is extremely doubtful. However, with respect to the training of the AEF, that is an insignificant question, since most of the technical instruction and training received by AEF individuals and units prepared the combat divisions to carry out only the limited, artillery-centered attacks that were the basis of an effective strategy of attrition. That is, the influence of the more conservative-minded Allied instructors was the opposite of what Pershing desired, but it appears to have adequately prepared the AEF to carry out the type of attacks that were best suited to American capabilities in 1918.

In their early engagements, most AEF divisions learned that they were fairly "weak" in their ability to carry out decisive operations. For the 2nd Division, the lesson was learned at Belleau Wood and confirmed at Soissons. However, at the same time that it experienced the bloody costs associated with realizing even modest operational achievements by using the "self-reliant infantry" approach, the 2nd Division discovered its ability to successfully execute more limited, artillery-centered, attritional attacks. This proficiency was further developed and demonstrated in the division's successful attacks at St. Mihiel, Blanc Mont, and in the Meuse-Argonne.

What emerged was a muted conflict between the divisional officers, who knew what their strengths and limitations were, and the Pershing-dominated GHQ, which proved reluctant to give up the glorious dream of the decisive, annihilating offensive and make the shift to waging the most effective war of attrition possible. The 2nd Division's commander, General Lejeune, claimed that he and his artillery brigade commander,

General Bowley, were in total agreement that "the artillery should be utilized to the limit of its capacity in order that victory might be achieved with the smallest possible loss of life among our assaulting troops."[22] It was hardly a description of Pershing's "self-reliant infantry" approach.

Operationally, one way that Lejeune's more conservative philosophy evidenced itself was in the maximum use of the preprogrammed rolling barrage. After the Aisne-Marne counteroffensive, Pershing released his *Combat Instructions* in which he stated that "the essential difference between open and trench warfare . . . is characterized by the presence or absence of the rolling barrage." Pershing knew that the rolling barrage was the most pronounced feature of the limited infantry attack and he longed to be rid of it. As a substitute, he offered the attacking divisions the German Army's tactic of breaking up artillery regiments into accompanying batteries and guns, in the hope of providing some supporting firepower to his infantrymen in the "open warfare" environment.[23]

The senior officers in the 2nd Division came up with a different method of supporting their attacking infantry, which was to extend the rolling barrage throughout the entire depth of the infantry's limited penetration. The rolling barrage for the St. Mihiel attack was maintained for a distance of more than eight kilometers, a clear indication that the senior officers in the division wanted to fight a more conservative, artillery-centered battle, rather than one based on aggressive advances of "self-reliant infantry," which were occasionally supported by a single 75mm gun here and there. Significantly, Lejeune and Bowley chose this artillery-centered method of attack when Pershing's plan had called for the extensive use of accompanying artillery. While there is no indication that the division received any criticism for this decision at St. Mihiel, General Lejeune was admonished by AEF inspectors for doing the same thing in the Meuse-Argonne attack. These facts encourage the view that the 2nd Division increasingly sought to fight limited, set-piece battles of attrition, while GHQ maintained a desire to fight a decisive war of annihilation.

There is yet more evidence that the field commanders (i.e., of divisions and corps) and the senior AEF and GHQ staff officers looked at their operations very differently. After the extremely successful First Army attack on 1 November in the Meuse-Argonne, in which the V Corps planned and executed a brilliant set-piece type of assault on the Barricourt Ridge, GHQ issued its assessment of the operation, entitled *Notes on Recent Operations No. 4*. Even though both divisions in V Corps (the 2nd and 89th Divisions) advanced well over five kilometers on the first day, and took all their objectives while suffering only modest casualties, the GHQ staff asserted that "the disadvantages of limited objectives were much in evidence." The assessment continued,

> As soon as assigned objectives were reached there almost invariably followed an immediate relaxation of all effort. Ordinarily, by this time the hostile resistance was furnished only by small rear guards. Usually there was still some daylight left . . . also a great part of the division concerned was comparatively fresh. *Under such circumstances pursuit to the limit of capacity of men and animals should have been the rule.* But the assignment of limited objectives immediately ended all thought of further efforts. (emphasis added)[24]

GHQ still had not learned that the continuation of the attack beyond the zone that was covered by a "sufficient superiority of fire" would almost surely have accelerated the decline in combat effectiveness, significantly increased the casualty rate, and achieved very little of consequence. At the end of the first day's advance, when the 2nd Division was consolidating its most forward positions, it was also bringing up its divisional guns so that future advances would have at least some concentrated artillery support.

As for V Corps' excellent fire-support plan for the initial attack, one that maximized the use of all artillery pieces in the entire corps throughout the depth of the whole first-day advance, the GHQ staff officers saw problems with that as well.

> The criticism of previous operations that too little use was made of the rifle and of all the infantry weapons must be repeated . . . the infantryman does not use his rifle enough. He is too prone to leave the whole matter of fire superiority to the artillery. This is not in accordance with American tradition or doctrine, and deprives the infantryman of the most efficient single aid to his advance—the rifle.[25]

In fact, the staff later made the assertion that it was the effective use of the rifle that had enabled some infantry units to take ground "at comparatively light cost." While the rifle may have been effective to a certain limited degree, it was the comprehensive fire-support plan, constructed around a heavy double- and triple-layered rolling barrage, that enabled most infantry units to advance without being shot to pieces.

The silent struggle between the operational units and GHQ was a competition between two different styles of warfare—the conservative, artillery-centered style of attrition versus the supposedly decisive infantry-focused style of annihilation. The combination of these two styles is what drove the distinctive pattern that emerged from nearly all AEF attacks. After the successful initial assault, which was often planned and carried out with set-piece precision by the combat corps and divisions, the divisions were rewarded by being ordered to immediately continue the advance in the vain hope of gaining some decisive breakthrough. The breakthroughs never came. Though divisions courageously attacked with greatly reduced firepower, after being given an inadequate amount of time to prepare the attack, they gained little, and suffered much.

This criticism does not fall exclusively on the American GHQ. As this investigation has shown, the same French Army whose conservative training instructors preached the doctrine of limited, set-piece, artillery-centered attacks to the new American divisions had its own faction of officers who pushed combat divisions to attack in extremely unfavorable situations. Marshal Foch and General Mangin were two such French officers whose devotion to the aggressive offensive, and whose willingness to order infantry attacks irrespective of the artillery support available, led to results similar to those caused by Pershing and his adherents in the AEF. The French Army also had officers like General Pétain, who were convinced that the more conservative, attritional approach was the most effective way to carry out offensives. While the French Army had fought long enough to have two well-developed, though contradictory, offensive strategies, the fledgling AEF did not fight long enough for the conservative, attritional approach to rise much above the level of the operational combat units. When the war ended, though there were senior commanders at the divisional and corps levels who were convinced that the strategy of attrition was needed to guide AEF operations, Pershing and his GHQ remained devoted to the strategy of annihilation. However, the inability of the AEF to implement its official annihilation-based strategy led it to fight in many ways a less efficient and more costly war than it could have fought.

Nevertheless, there were units in the AEF, such as the 2nd Division, that had clearly demonstrated the will and the ability to effectively carry out limited, artillery-centered, set-piece attacks. Therefore, the revisionists broad-brush criticism of the AEF as a collection of basically incompetent units should not be accepted as the full story. There is little doubt that the 2nd Division was among the best that the AEF had to offer, but there is no reason to believe that it was vastly superior to all other American divisions.[26] The other three "pioneer" divisions, the 1st, 26th and 42nd Divisions, which were the first American divisions to reach France, all received as much or more training than the 2nd Division. Together these four divisions suffered over 30 percent of all AEF casualties and accounted for 35 percent of all German prisoners taken during the war. If one adds just two other active divisions, the 3rd and 28th Divisions, we can account for over 45 percent of all AEF casualties and 40 percent of prisoners taken.[27] Considering the short duration of the AEF's war experience, these few divisions accounted for a disproportionate amount of the total combat effort. While it is probable that the 2nd Division was much superior to divisions that saw very little combat, it is also likely that some or all of the divisions that carried the bulk of the combat load approximately matched the 2nd Division's level of expertise. This suggests that the AEF was in some ways a more effective combat force than the revisionists have claimed.

What this analysis has identified as a fundamental weakness of the AEF was the great variance between the style of warfare advocated by Pershing, GHQ, and the War Department and the style that the combat divisions and corps increasingly desired and attempted to fight. The AEF High Command strove to execute a style of warfare that was matched more to the tradition of the American way of war than to the capabilities of the units that were tasked to do the fighting. The first result was an excessively steep learning curve for the new divisions, as they not only had to overcome their natural inexperience but also had to disabuse themselves of the notion that they could effectively fight in the style preached by their most senior officers. Those that fought long enough, like the 2nd Division, eventually traded that aggressive infantry-based style for the more conservative artillery-centered style, and they proved that they could do battle under the latter's terms with good effect. The second and more tragic result of the High Command's poorly matched style of warfare was that the offensives that were based on it squandered the genuinely effective capabilities of the best AEF divisions. The AEF High Command's adherence to what it perceived to be the traditional American style of war led to excessive casualties, often with little to show for them.

EPILOGUE

The doctrinal conflict that was brewing in the AEF did not disappear when the war ended in 1918. When the AEF was sent back to the United States and demobilized, officers like General Summerall and Colonel Lanza carried their beliefs about the appropriate style of warfare back with them. While they advocated a military force that was organized, trained, and equipped to fight battles based on the massive use of heavy firepower, others in the U.S. Army maintained their allegiance to the traditional style based on flexibility and rapid maneuver.

As the United States followed its historical pattern of fully demobilizing its war-time army and returning to a small peace-time standing force, the U.S. Army unsuccessfully struggled to gain the resources necessary to put the two competing doctrines to the test on the training ground. One historian of the American military during the interwar period has stated that the army "never really made a choice between a doctrine of applying mass and overwhelming power to crush the enemy and a doctrine which emphasized mobility and maneuver."[28] This unsettled issue, which had its roots in the AEF's employment of artillery, was to have a negative impact on American military preparedness throughout the interwar period, and on the effectiveness of the U.S. Army in World War II.[29]

NOTES

1. Lanza, "Artillery Support," pp. 72, 84–85.

2. Of the more than 255,000 Americans killed and wounded in the war, over 56 percent were caused by shell and shrapnel and nearly 32 percent by gas shells. Only about 9 percent of casualties were the result of rifle and machine-gun fire. Lanza, "Counterbattery in the A.E.F.," p. 466, and ABMC, *American Armies*, p. 515.

3. John B. Anderson, "Are We Justified in Discarding 'Pre-War' Methods of Training?," *FAJ* (April–June, 1919), p. 225.

4. Pershing received a Pulitzer Prize for his memoirs, entitled *My Experiences in the World War*. Paschall, *Defeat of Imperial Germany*, p. 229.

5. Letter, Pershing to Major General Ernest Hinds, Chief of Artillery, AEF, 27 March 1919, in "Report of Operations of the Chief of Artillery, AEF," 10 May 1919.

6. Smythe, *Pershing*, p. 129. On 28 July 1918, Pershing wrote to the American secretary of war and boldly proclaimed, "The fact is that our officers and men are far and away superior to the tired Europeans." Pershing, *My Experiences*, II: 189.

7. John J. Pershing, "Final Report," 1 September 1919, p. 22; Pershing, "Preliminary Report," 16 January 1919, pp. 6–7. Both in *USAWW*, vol. 12.

8. Pershing claimed that it was the "spirit of determination animating every American soldier that made it impossible for the enemy to maintain the struggle until 1919." See Pershing, "Final Report," *USAWW*, 12: 44.

9. Rainey, "The Questionable Training of the AEF in World War I," *Parameters*, Journal of the U.S. Army War College (Winter, 1992–93), p. 100.

10. Braim, *Test of Battle*, p. 153.

11. Trask, *Coalition Warmaking*, pp. 141, 175.

12. Ibid., pp. 71–73, 92–93.

13. Fewer than 53,000 Americans suffered combat-related deaths. For comparison, France suffered approximately 1,400,000 war deaths and the British Empire nearly one million deaths. American figures taken from ABMC, *American Armies*, p. 515; other casualty figures come from Paschall, *Defeat of Imperial Germany*, p. 224.

14. Report of Major General Summerall to Deputy Chief of Staff, AEF, 4 January 1919, Subject: "Artillery," in V Corps Historical File, Entry 1118, RG 120, NA.

15. Ibid.

16. For example, the U.S. 27th and 30th Divisions trained with the British, fought under British control near Ypres in August and September 1918, and again during the British Fourth Army offensive in late September and October 1918. The U.S. 33rd Division fought with the BEF at the battle of Hamel on 4 July 1918, and later served admirably from 9–19 August in the British offensive against the Amiens salient.

17. Conrad Lanza, "Note on the Artillery in the Battle of Buzancy, November 1, 1918," *FAJ* (March–April, 1932), p. 160.

18. Lanza, "Artillery Support," pp. 71–72, 83.

19. See Summerall, "Recent Operations from the Standpoint of Employment of Artillery with Particular Reference to Co-operation between the Infantry and Artillery," Lecture to ACAS, 16 December 1918, in Folder 14, ACAS File, Entry 371, and "Comments by the Corps Commander upon the Operations of the Fifth Army Corps," V Corps Historical File, Entry 1118, both RG 120, NA.

20. Russell F. Weigley, *The American Way of War* (Bloomington: Indiana University Press, 1973), pp. xvii–xxiii.

21. Ibid., p. xxii.

22. Lejeune, *Reminiscences*, p. 383.

23. Pershing, *Combat Instructions*, p. 1.

24. GHQ, AEF, *Notes on Recent Operations No. 4* (France, 22 November 1918), p. 1.

25. GHQ, AEF, *Notes on Recent Operations No. 4* (France, 22 November 1918), p. 1.

26. Paul Braim does rate the 2nd Division as the best overall division in the AEF, followed by the 1st, 3rd, and 42nd. See Braim, p. 149.

27. These figures come from ABMC, *American Armies*, p. 517. The four "pioneer divisions" constituted just 14 percent of the twenty-nine AEF divisions that saw combat.

28. Ronald Spector, "The Military Effectiveness of the US Armed Forces, 1919–39," in Allan R. Millett and Williamson Murray, eds., *Military Effectiveness*, Volume II: *The Interwar Period* (Boston: Allen and Unwin, 1988), p. 90.

29. Spector cites Russell Weigley's assertion that while the U.S. Army's tactical concepts in World War II emphasized the use of overwhelming firepower "to crush the enemy in protracted battle," the army's weapons and equipment were better suited to the style of warfare based on mobility and maneuver. See Spector, "Effectiveness of US Armed Forces," p. 90.

Appendix A: Artillery Used by the AEF

Weapon (transport)	Producer	Projectiles	Weight of Explosive	Maximum Range
Divisional Artillery				
75mm Gun (horse)	France	HE Shell 1917 "D" Shell Shrapnel	0.775 kg	8,000 meters 11,000 meters 8,500 meters
155mm How. (primarily horse, some motorized)	France	HE Shell Shrapnel Elongated Shell	4.500 kg 10.200 kg	11,200 meters 9,500 meters
58mm Trench Mortar (horse)	France	Hollow Tail Bomb	6.150 kg	1,500 meters
6 inch Newton Trench Mortar (horse)	Great Britain	Bomb	10 lbs	1420 meters

Corps and Army Artillery				
155mm Gun (tractor)	France	Elongated Shell	7.150 kg	16,200 meters
8 inch How. (tractor/horse)	Great Britain	HE Shell	29.69 lbs	11,600 meters
9.2 inch How. (tractor)	Great Britain	HE Shell	34.00 lbs	11,600 meters
240mm Trench Mortar (horse)	France	Bomb	42.00 kgs	2,200 meters
9.45 inch Newton Mortar (horse)	Great Britain	Bomb	56.00 lbs	1,050 meters

Taken from the table "Artillery Used by U.S. Army in France," G-S Schools Files, Entry 362, RG 120, NA

Appendix B: 2nd Division Organizational Chart

Unit	Officers / Men	Weapons	Commander Rank	Remarks
Division Staff	45 / 200	N/A	Colonel	
Infantry Brigade	258 / 8211	6457-rifles 780-rifle grenades 384-automatic rifles 96-machine guns 12-3" Stokes mortars 6 - 37mm guns	Brigadier General	2 per division
Machine gun Battalion	27 / 730	64-machine guns 40-rifles / 717-pistols	Major	1 per inf. brigade
Infantry Regiment	112 / 3720	3200-rifles 390-rifle grenades 192-automatic rifles 16-machine guns 6-3" Stokes mortars 3 - 37mm guns	Colonel	2 per inf. brigade (4 per division)
Infantry Battalion	27 / 1000	940-rifles 120-rifle grenades 64-automatic rifles	Major	3 per inf. regiment (6 per brigade/12 per division)
Rifle Company	6 / 250	235 - rifles 73 - pistols 30 - rifle grenades 16 - automatic rifles	Captain	4 per inf. battalion (12 per regiment/ 24 per brigade/ 48 per division)
Rifle Platoon	1 / 58	54-rifles / 16-pistols 6 - rifle grenades 4 - automatic rifles	First Lieutenant	4 per company
Machine gun Company	6 / 172	16-machine guns 178-pistols	Captain	1 per inf. regiment

Artillery Brigade	214 / 4841	24-155mm howitzers 48-75mm guns 12-6"/58mm mortars	Brigadier General	1 per division
75mm Regiment	64 / 1501	24- 75mm guns	Colonel	2 per division
75mm Battalion	19 / 582	12-75mm guns	Major	2 per regiment
75mm Battery	5 / 194	4 -75mm guns	Captain	3 per battalion
155mm Regiment	65 / 1727	24-155mm howitzers	Colonel	1 per division
155mm Battalion	16 / 456	8-155mm howitzers	Major	3 per regiment
155mm Battery	5 / 228	4-155mm howitzers	Captain	2 per battalion
Trench Mortar Battery	5 / 172	12-6" Newton or 58mm Tr. Mortars	Captain	1 per division
Signal Battalion	15 / 473	473 pistols	Major	1 per division
Engineer Regiment	51 / 1646	1487 rifles/179 pistols	Colonel	1 per division
Supply Trains	122 / 3121	NA	various ranks	total for all trains
		Division Totals		
	Officers/Men	Weapons	Commander Rank	Remarks
Combat Division	979 / 27,080 total combatant 847 / 25,553	17,666 total rifles 1,560-grenade launchers 768- automatic rifles 224- machine guns 72- artillery pieces 36-anti-aircraft guns 26- Trench mortars 12- 37mm guns	Major General	3936 horses 2700 mules 577 motor trucks 319 motorcycles 122 motor cars 64 tractors

Taken from *United States Army in the World War,* vol. 1, pp. 339–88.

Bibliography

PRIMARY SOURCES

I. The National Archives, Washington, D.C.

Record Group 120: Records of the American Expeditionary Forces (World War I), 1917–1923.

Entry 22: Reports of the Commander-in-Chief, AEF, Staff Sections, and Services Relating to the History of the US Army in Europe During World War I.

Entry 270: G-3 Special Operations Reports, General Staff, AEF.

Entry 362: Records of the American General Staff College, G-5 Schools File, General Staff, AEF.

Entry 371: Records of the Army Center of Artillery Studies.

Entry 443: Reports and Other Records Regarding the History of AEF Units, Division Skeleton Histories.

Entry 590: Artillery Files, Inspector General Reports, AEF.

Entry 692: Records of the Valdahon Artillery School.

Entry 990: I Corps Historical File.

Entry 1118: V Corps Historical File.

Entry 1139: Correspondence of the V Corps Artillery Chief.

Entry 1241: 2d Division Historical File, Records of the 1st–20th, 26th–42d, and 76th–103rd Divisions.

Entry 1243: 2nd Trench Mortar Battery File, Records of the 1st–29th, 101st–117th, and 301st–317th Trench Mortar Artillery Batteries.

Entry 1246: 2nd Field Artillery Brigade Historical File, Records of the 1st–24th, 83rd, 151st–167th, and 170th–172nd FA Brigades.

Record Group 165: Records of the War Department and Special Staffs.

Entry 318: French Military Records Relating to World War I, 1917–1919.

II. U.S. Army Military History Institute, Carlisle Barracks, PA

Army Service School. *Instructions for the Training of Divisions for Offensive Action.* Fort Leavenworth, KS: Army Service Schools Press, 1917.

Cone, Anthony D. *E Battery Goes to War: Being Extracts from the Wartime Diary of First Sgt. Anthony D. Cone, 15th Field Artillery, 2 Division.* Washington, D.C.: Raymond E. Lee, 1929.

GHQ, AEF. Notes on Recent Operations, No. 1. France, 7 August, 1918.

———. Notes on Recent Operations, No. 3. France, 12 October, 1918.

———. Notes on Recent Operations, No. 4. France, 22 November, 1918.

Lanza, Conrad H., Colonel. *The Army Artillery, First Army.* 2 vols. Typescript copy of unit history, written for the War Department. No date.

Pershing, John J. *Combat Instructions.* AEF Document No. 1348. France: GHQ, AEF, 5 September, 1918.

U.S. Army, the Chief of Field Artillery. *The School of the Battery Commander: 75m/m gun and 155m/m howitzer.* From the pamphlet at the Saumer Artillery School, France. Fort Sill, OK: November 1918.

U.S. Army, 2nd Division. *Records of the Second Division (Regular).* 10 vols. (some in several parts). Washington, D.C.: Army War College, 1924.

U.S. War Department. *Field Service Regulations: United States Army, 1914, Corrected to July 31, 1918.* Washington, D.C.: U.S. Government Printing Office, 1918.

———. *Provisional Field and Service Regulations for Field Artillery (Horse and Light), 1916, Corrected to April 15, 1917.* New York: Military Publishing Company, 1917.

Vencill, Ralph K. *From Robinson to the Rhine: Battery "A" 17th Field Artillery, 1917–1918–1919.* Kindt and Meinardus, 1919.

World War I Manuscript Collection: 2nd Division boxes.

III. The Colonel Robert R. McCormick Research Center, First Division Museum at Cantigny

Butler, Alban B. *Happy Days!* Washington, D.C.: Society of the First Division, A.E.F., 1928.

U.S. Army, 1st Division. *World War Records of the First Division A.E.F. (Regular).* 25 vols. Washington, D.C.: Army War College, 1930.

IV. The American Official History of the U.S. Army in World War I

U.S. Department of the Army, Historical Division. *United States Army in the World War 1917–1919.* 17 vols. Washington, D.C.: U.S. Government Printing Office, 1948.

V. Journal Articles

Anderson, John B. "Are We Justified in Discarding 'Pre-War' Methods of Training?" *Field Artillery Journal* (April–June, 1919): 222–30.

"Current Field Artillery Notes." *Field Artillery Journal* (April–June, 1917): 198–204.

Department of Gunnery, School of Fire for Field Artillery. "American Drill Regulations and 'Artillery Firing'." *Field Artillery Journal* (July–September, 1918): 363–69.

Editorial Department. *Field Artillery Journal* (January–March, 1917): 71–74.

Fleming, A. S., Colonel. "The Mission of the School of Fire for Field Artillery: Address to Incoming Class by Commandant, Col. A. S. Fleming." *Field Artillery Journal* (October–December, 1917): 383–90.

"German Artillery." *Field Artillery Journal* (October–December, 1918): 578–85.

Irwin, George LeR., Brigadier General. "Notes on the Training and Handling of Divisional Artillery in France." *Field Artillery Journal* (November–December, 1919): 489–507.

Lanza, Conrad H., Colonel. "Counterbattery in the A.E.F." *Field Artillery Journal* (July–September, 1936): 454–79.

———. "Note on the Artillery in the Battle of Buzancy, November 1, 1918." *Field Artillery Journal* (March–April, 1932): 157–60.

———. "The Artillery Support of the Infantry in the A.E.F." *Field Artillery Journal* (January–March, 1936): 62–85.

Locke, E. M., Major. "Artillery in Europe." *Field Artillery Journal* (July–September, 1917): 294–301.

"Measures Taken by the German Artillery to Carry Out Preparations for Attack Without Betraying the Intentions of the Command." *Field Artillery Journal* (July–September, 1918): 504–12.

"New Field Artillery Classification." *Field Artillery Journal* (January–March, 1917): 25–30.

"Notes on Artillery." *Field Artillery Journal* (April–June, 1917): 164–97.

"Organization of a Rolling Barrage in the German Army: Translation of a German Document." *Field Artillery Journal* (July–September, 1918): 417–21.

Reilly, Henry J., Colonel. "Fontainebleau in War Time." *Field Artillery Journal* (April–June, 1917): 109–18.

"The Scientific Preparation of Fire in the German Army." *Field Artillery Journal* (October–December, 1918): 527–34.

VI. Published Memoirs and Other First-Hand Accounts

Brannen, Carl Andrew. *Over There: A Marine in the Great War*. College Station: Texas A&M Press, 1996.

Bullard, Robert L. *American Soldiers Also Fought*. New York: Longmans, Greene and Co., 1936.

Chambrun, Jaques and Charles Marenches. *The American Army in the European Conflict*. New York: MacMillan, 1919.

Harbord, James G. *The American Army in France, 1917–1919*. Boston: Little, Brown, 1936.

———. *Leaves from a War Diary*. New York: Dodd, Mead, 1925.

Lejeune, John A. *The Reminiscences of a Marine*. Philadelphia: Dorrance and Company, 1930.

Liggett, Hunter. *A.E.F.: Ten Years Ago in France*. New York: Dodd, Mead, 1928.

———. *Commanding an American Army*. New York: Houghton Mifflin, 1925.

Mackin, Elton E. *Suddenly We Didn't Want to Die: Memoirs of a World War I Marine*. Novato, CA: Presidio Press, 1993.

March, Peyton C. *The Nation at War*. Garden City, NY: Doubleday, Doran, 1932.

Pershing, John J. *My Experiences in the World War*. 2 vols. New York: Frederick A. Stokes Company, 1931.

Snow, William J. *Signposts of Experience*. Washington, D.C.: U.S. Field Artillery Association, 1941.

SECONDARY SOURCES

I. Books

American Battle Monuments Commission. *American Armies and Battlefields in Europe: A History, Guide, and Reference Book*. Washington, D.C.: U.S. Government Printing Office, 1995 (Center for Military History reprint), 1938.

Asprey, Robert B. *At Belleau Wood*. New York: Putnam's Sons, 1965.

Bailey, J. B. A. *Field Artillery and Firepower*. Oxford: The Military Press, 1989.

Bidwell, Shelford, and Dominick Graham. *Fire-Power: British Army Weapons and Theories of War, 1904–1945*. London: Allen and Unwin, 1982.

Braim, Paul F. *The Test of Battle: The American Expeditionary Forces in the Meuse-Argonne Campaign*. Newark: University of Delaware Press, 1987.

Bruce, Robert. *Machine Guns of World War I*. London: Windrow and Greene, 1997.

Cochrane, Rexmond C. *Gas Warfare at Belleau Wood, June 1918*. Study #1, U.S. Army Chemical Corps Historical Studies: Gas Warfare in World War I. Army Chemical Center, Maryland, 1957.

———. *The Use of Gas in the Meuse-Argonne Campaign, September–November 1918*. Study # 10, U.S. Army Chemical Corps Historical Studies: Gas Warfare in World War I. Army Chemical Center, Maryland, 1958.

Coffman, Edward M. *The War to End All Wars*. New York: Oxford University Press, 1968.

Comparato, Frank. *Age of the Great Guns*. Harrisburg, PA: The Stackpole Company, 1965.

Dastrup, Boyd S. *King of Battle: A Branch History of the U.S. Army's Field Artillery*. Ft. Monroe, VA: U.S. Army Training and Doctrine Command, 1992.

Demaison, Gerard, and Yves Buffetaut. *Honour Bound: The Chauchat Machine Rifle*. Cobourg, ONT: Collector Grade Publications, 1995.

DeWeerd, Harvey A. *President Wilson Fights His War: World War I and the American Intervention*. New York: Macmillan, 1968.

Downey, Fairfax. *The Sound of the Guns: The Story of American Artillery*. New York: David McKay, 1955.

Gudmundsson, Bruce I. *On Artillery*. Westport, CT: Praeger, 1993.

———. *Stormtroop Tactics: Innovation in the German Army, 1914–1918.* New York: Praeger, 1989.

Heller, Charles E. "Chemical Warfare in World War I: The American Experience, 1917–1918." Leavenworth Papers No. 10. Fort Leavenworth, KS: U.S. Army Combat Studies Institute, September, 1984.

Herwig, Holger H. *The First World War: Germany and Austria-Hungary 1914–1918.* London: Arnold, 1997.

Hogg, Ivan. *The Guns, 1914–1918.* New York: Ballantine Books, 1971.

Millett, Allan R. *The General: Robert L. Bullard and Officership in the United States Army, 1881–1925.* Westport, CT: Greenwood Press, 1975.

———. *Semper Fidelis: The History of the United States Marine Corps.* New York: Macmillan, 1980.

Nenninger, Timothy K. *The Leavenworth Schools and the Old Army.* Westport, CT: Greenwood Press, 1978.

Paschall, Rod. *The Defeat of Imperial Germany, 1917–1918.* New York: Da Capo Press, 1994.

Schreiber, Shane B. *Shock Army of the British Empire: The Canadian Corps in the Last 100 Days of the Great War.* Westport, CT: Praeger, 1997.

Smythe, Donald. *Guerrilla Warrior: The Early Life of John J. Pershing.* New York: Charles Scribner's Sons, 1973.

———. *Pershing: General of the Armies.* Bloomington: Indiana University Press, 1986.

Spaulding, Oliver and John W. Wright. *The Second Division, American Expeditionary Force in France, 1917–1919.* New York: The Hillman Press, 1937.

Trask, David F. *The AEF and Coalition Warmaking, 1917–1918.* Lawrence: University Press of Kansas, 1993.

Travers, Timothy H. E. *How the War was Won: Command and Technology in the British Army on the Western Front, 1917–1918.* London: Routledge, 1992.

U.S. War Department. *Order of Battle of the United States Land Forces in the World War.* 3 vols. Washington, D.C.: U.S. Government Printing Office, 1995 (Center for Military History reprint), 1937.

Weigley, Russell F. *The American Way of War.* Bloomington: Indiana University Press, 1973.

———. *History of the United States Army.* Bloomington: Indiana University Press, 1967.

Zabecki, David T. *Steel Wind: Colonel Georg Bruchmüller and the Birth of Modern Artillery.* Westport, CT: Praeger, 1994.

II. Contemporary Articles

Jannette, Alexander T. "Mass Fire in WWI." *Field Artillery Journal* (May–June, 1975): 40–45.

Kennedy, Paul. "Britain in the First World War." In *Military Effectiveness.* Volume I: *The First World War,* edited by Allan R. Millett and Williamson Murray. Boston: Allen and Unwin, 1988, pp. 31–79.

Nenninger, Timothy K. "American Military Effectiveness in the First World War." In *Military Effectiveness.* Volume I: *The First World War,* edited by

Allan R. Millett and Williamson Murray. Boston: Allen and Unwin, 1988, pp. 116–56.

———. "Tactical Dysfunction in the AEF, 1917–1918." *Military Affairs* (October, 1987): 177–81.

Rainey, James W. "Ambivalent Warfare: The Tactical Doctrine of the AEF in World War I." *Parameters*, Journal of the U.S. Army War College (September 1983): 34–46.

———. "The Questionable Training of the AEF in World War I." *Parameters*, Journal of the U.S. Army War College (Winter, 1992–93): 89–103.

Spector, Ronald. "The Military Effectiveness of the US Armed Forces, 1919–1939." In *Military Effectiveness*. Volume II: *The Interwar Period*, edited by Allan R. Millett and Williamson Murray. Boston: Allen and Unwin, 1988, pp. 70–97.

III. Unpublished Theses

Pierce, Richard Lee. "A Maximum of Support: The Development of U.S. Army Field Artillery Doctrine in World War I." Master's thesis, Ohio State University, 1983.

Sinnema, Robert J. "General Mangin and the Tenth Army, 1918." Master's thesis, University of Calgary, 1995.

Stebbins, Steven A. "Indirect Fire: The Challenge and Response in the US Army, 1907–1917." Master's thesis, University of North Carolina, 1993.

Index

Note: **Boldface** numbers indicate illustrations.

About the Author

MARK E. GROTELUESCHEN is Assistant Professor of History at the United States Air Force Academy.